ONE SMALL STEP

The inside story of space exploration

David Whitehouse

Quercus

Contents

Liftoff of the *Apollo 17* mission - the last to land on the Moon.

Introduction

An American president lays down a challenge to his nation that he will never live to see fulfilled. A Russian despot curses his rocket engineers and urges them to greater effort. A giant rocket explodes in a deadly fireball seconds after lift-off, ending the lives of those on board and the dreams of those watching on the ground. One moonwalker takes mankind's first steps on another world, while another writes his daughter's name in the lunar dust. A German SS soldier dreams of space, while at the same time a prisoner digs a grave in a Soviet death camp, never thinking that he will ever see space travel – let alone be one of its key pioneers. A crew of astronauts come close to a lonely death on their journey to the Moon. Another crew, at another time, know they will never reach home again.

These are just a few of the multitude of momentous events that have come to symbolize our enduring fascination with space travel and space exploration. When do such events become legends? When does our history turn to something more than merely moments in time? How long does it take for us to appreciate the true significance of the times we have lived through? Without doubt, when the history of the 20th century is written, one of its the major landmarks will be the journey into space. In many years to come, when much of today's modern history will be footnotes, *Sputnik*, Gagarin, Armstrong and his *'one small step'* will still be headlines. Across the growing centuries future historians will look upon the 20th century as the time when things changed forever, the time when mankind left its home planet and ventured out but a short distance into space. When, for a while, we achieved greatness.

Strange then that it is a story so few know in any detail. In schools we teach the voyages of Columbus, Vasco da Gama, Marco Polo and Amundsen, but seldom the greater voyages of Gagarin, Borman, Lovell or Anders, or even those of Armstrong, Aldrin and Collins.

There are many threads that can be drawn through the story of the astronauts, and each time the story is told, it is a slightly different weave. This is one story about the great things spacemen and space women have done for us; of the way they wrote history. Dreams and tragedies, pride and mourning, conflict and immortality – all are intertwined in the story of mankind's steps in space. It is not just a technological tale of rockets, satellites and spaceships. It is much more a story of special human beings overcoming all obstacles and setbacks to fulfil their dreams and the dreams of others. It is, I believe, our greatest story, and it is best told in the words of those who were there.

Over the years I have been fortunate enough to meet many astronauts, cosmonauts and others involved in this great adventure, and they have been kind enough to share their stories with me. I am also indebted to NASA for its extensive history programme that is a rich source of interviews, analysis and technical data.

David Whitehouse

'Soviet rockets must conquer space!'

REALIZING THE DREAM

WERNHER VON BRAUN AND SERGEI KOROLEV
1903–1957

In the 20th century it finally began to seem that Mankind's long-held vision of travelling in space could become a reality. Yet despite the peaceful ambitions of the early pioneers of rocketry and space flight, the technology needed for such an event was developed initially to produce weapons of war.

First there were the dreamers, then came the practical men and women, and finally the voyagers themselves. The origin of the space programmes of the 20th century can perhaps be traced back to the late 1800s and the ideas of a deaf, self-educated Russian school teacher called Konstantin Eduardovich Tsiolkovsky. Born in 1857, his writings, frequently visionary if sometimes far-fetched, put substance to mankind's nascent dreams of escaping from our planet and journeying into the cosmos. He wrote of multistage rocket boosters and space stations orbiting the Earth. It was Tsiolkovsky, along with the later generation of rocket pioneers, the American Robert Goddard and the German Hermann Oberth, who prepared the way for what followed. But while all three dreamed of space travel, only Tsiolkovsky thought it would never come to pass.

Driving the Space Age Forward

Two men have become synonymous with the start of the Space Age. Both were fortunate to survive the Second World War. Separated by the Iron Curtain, they never met, but they dreamed the same dreams, were both victims and beneficiaries of politics and both achieved remarkable things.

The first of these great pioneers was Wernher von Braun, who was born in Germany just before the First World War. The von Braun family had been famous since 1245, when they defended Prussia from Mongol invasion. Wernher von Braun had showed an interest in rockets from an early age; when Hermann Oberth, Germany's foremost rocket scientist, had written a book entitled *Die Rakete zu den Planetenräumen* (*The Rocket into Planetary Space*), in which he describes a rocket designed to go to the Moon. A young von Braun read it and was captivated.

While the young von Braun dreamed, a Russian called Fredrikh Tsander would greet his fellow workers with the phrase: 'To Mars! Onward to Mars!' Tsander was born in 1887 in Riga, Latvia. By his late twenties

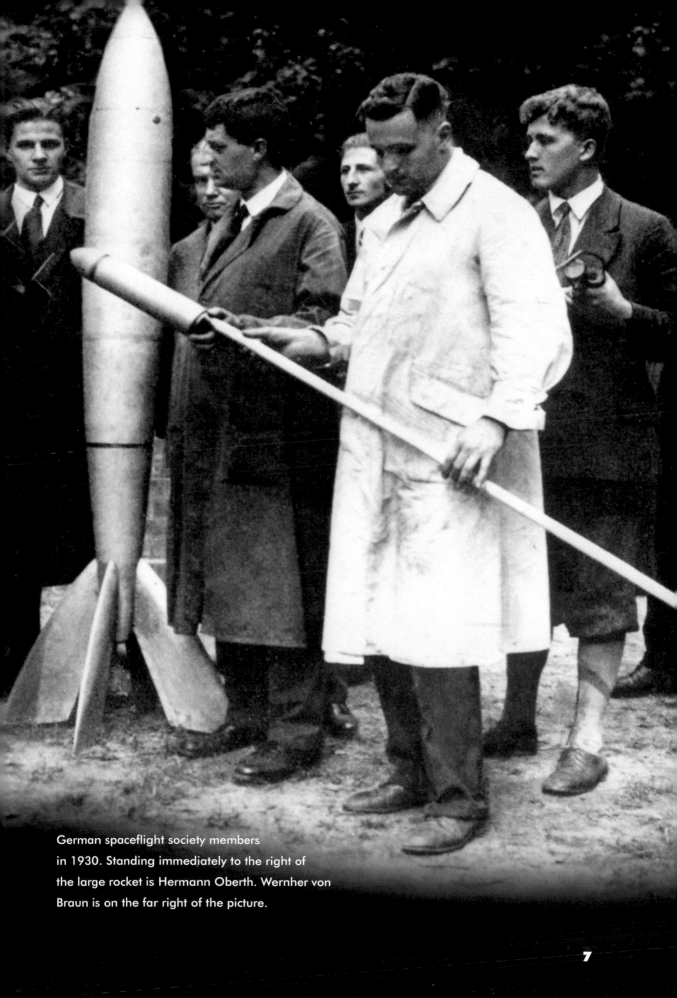

German spaceflight society members in 1930. Standing immediately to the right of the large rocket is Hermann Oberth. Wernher von Braun is on the far right of the picture.

Sergei Korolev in 1954 with a dog that had been sent in a rocket to an altitude of 62 miles (100 km).

he had decided he wanted to make a journey into space. In 1924 he published his landmark work entitled *Flight to Other Planets*, in which he described the design of rocket engines and made calculations for interplanetary trajectories. But perhaps his most significant contribution was his untiring popularization of spaceflight through his lecturing on the topic across the Soviet Union. Small rocketry societies were formed in Moscow and Leningrad by enthusiasts who wanted to build short-range, liquid-fuelled rockets. Among them was the other person whose story is synonymous with spaceflight: Sergei Pavlovich Korolev.

Korolev, who was born in 1906, in the town of Zhitomir in the Ukraine, did not go to school until he was 14. He dreamed of flight, voraciously reading the exploits of aviation pioneers, and at 17 he joined a glider club in Odessa. Two years later he enrolled in the Moscow Higher Technical School in the Department of Aerodynamics, where he came into contact with famous Soviet aeronautical designers such as Andrei Tupolev. Upon graduation, he was employed as an engineer to work on aircraft engine design at the Menzhinski Central Design Bureau. After a few months he was transferred to the prestigious Central Aerohydrodynamics Institute in Moscow.

A Momentous Meeting

It appears that Korolev would have had a sound career ahead of him designing aircraft. But then he met Fredrikh Tsander. Tsander had tried to secure government support for his rocketry experiments, but had met with no success. Almost in desperation he placed an advertisement in a Moscow newspaper inviting contact from anyone interested in 'interplanetary communications'. Over 150 people responded. So it was that in July 1931 some of them formed the Bureau for the Investigation of Reactive Engines and Reactive Flight, later changing its name to the Group for the Investigation of Reactive Engines and Reactive Flight (GIRD). Korolev was a key member.

Under Tsander's leadership, GIRD held public lectures and carried out small experiments in a wine cellar on Sadovo-Spasskiy Street in Moscow. Soon Korolev replaced the ailing Tsander as leader and, using his administrative flair, established four research groups to study different problems associated with rocketry. Money started to flow from the government, and by the late summer of 1933 they were able to launch the Soviet Union's first liquid-fuelled rocket powered by jellied petroleum burning in liquid oxygen. After two failures the third attempt soared to 400 metres (1,312 ft). Korolev wrote:

From this moment Soviet rockets should start flying above the Union of Republics. Soviet rockets must conquer space!

Tsander did not live to see the triumph. Five months earlier, exhausted by overwork, he had contracted typhus and died.

The Soviet government was impressed by the rocket launch, and soon Korolev and his colleagues were working for it. The government was already funding another small research group investigating solid-fuelled rockets for military use, led by a young engineer called Valentin Petrovich Glushko. He had been inspired by the works of Jules Verne, and at 15 he had written a letter to Tsiolkovsky. Just three years later in 1924, when still only 18, he had published an article in the popular press entitled *Conquest of the Moon by the Earth*. Glushko and Korolev became friends, but that was not to last. Their difficult relationship was to be at the heart of the Soviet space effort, becoming both its greatest strength and weakness.

'Soviet rockets must conquer space!'

SERGEI KOROLEV

Military Backing

In Hitler's Germany a young Wernher von Braun showed his rocketry ideas to Colonel Karl Becker, chief of ballistics and ammunition of the Reichswehr (National Defence). Becker responded:

> *We are greatly interested in rocketry, but there are a number of defects in the manner in which your organization is going about development. For our purposes, there is too much showmanship. You would do better to concentrate on scientific data than to fire toy rockets.*

Wernher von Braun was interested in using rockets for space flight, but Becker wanted a long-range missile. Von Braun nevertheless joined the army and worked under Captain Walter Dornberger on liquid-fuelled rocket engines, saying later:

> *We needed money for our experiments, and since the army was willing to give us help, we didn't worry overmuch about the consequences in the distant future. We were interested in one thing, the exploration of space.*

Because the army was the only organization developing rockets in Germany, and also to advance his own career, von Braun joined first the Nazi Party and then the Waffen-SS. He told his fellow amateur rocket enthusiasts that he had been conscripted, and then started to lie to them about what he was really doing. In the meantime Dornberger decided that a quiet and isolated place was needed for the rocket tests, so the programme was moved to a small fishing village called Peenemünde on the Baltic Sea.

The Rule of Terror

The promise and potential of the Soviet rocketry effort was cut short abruptly in 1937 when Joseph Stalin's purges reached their inhuman climax. His plan – the complete terrorization of society – was put into effect by the hated People's Commissariat for Internal Affairs (NKVD, the secret police force responsible for political repression). An entire generation, and more, of Soviet society was murdered. No one was safe, and

1903

May First publication of Russian Konstantin Tsiolkovsky's work *The Exploration of the World Space with Jet Propulsion Instruments*

1909

American Robert Goddard starts research in the field of rocket dynamics

1917

5 January The Smithsonian Institution awards a $5,000 grant to Robert Goddard to conduct rocket research of the upper atmosphere

1923

In Germany, Hermann Oberth publishes *The Rocket into Interplanetary Space*

1928

11 June The world's first aircraft powered by a rocket engine completes a flight in Germany

1930

18 September Russian Fredrikh Tsander conducts the first tests of liquid-fuelled engines

1932

14 July The Soviet government begins sponsoring Moscow-based Group for the Investigation of Reactive Engines and Reactive Flight (GIRD)

continued on p.14

nor was there any defence. People simply disappeared, often picked up off the street for completely arbitrary reasons. Millions faced the threat of execution or being sent to labour camps. Terrified people became informants simply to survive – among them Valentin Glushko.

By the end of 1937 the NKVD had Korolev and Glushko in their sights. Glushko was arrested first. Inevitably, the NKVD denounced Korolev and he was thrown into the Lubyanka. Shortly afterwards, following severe torture, he 'confessed' and was fortunate not to be shot. Instead, he found himself in a cattle truck being taken to the Kolyma death camp in Siberia.

Two chance events saved his life. A close friend, the famous pilot Valentina Grizodubova, joined forces with another famous Soviet aviator, Mikhail Gromov, and with Korolev's mother to write a letter to the Central Committee of the Communist Party requesting a review of his case. It reached the office of Nikolai Yezhov, chairman of the NKVD, and his successor, the tyrannical Lavrenti Beriya. Prosecutor Vasily Ulrikh also wrote to the NKVD to protest at Korolev's sentence. Beriya thought he could use Korolev's case to demonstrate his powers of leniency. So at a special meeting of the Plenum of the High Court the NKVD agreed to Ulrikh's request and altered the charge from a 'member of an anti-Soviet counterrevolutionary organization' to the less serious 'saboteur of military technology', and ordered a new trial.

For Korolev, working as a grave digger in a gold mine off the Kolyma river, it was almost too late. Among all the gulags Kolyma was the most brutal, claiming the lives of between two and three million people from overwork, famine, cruelty and the harsh Arctic climate. Eventually Korolev was found at Kolyma before his inevitable death and put on a train back to Moscow. Of the 600 individuals who had been at the camp when he had arrived, only 200 were still alive when he left. Soon after, under Beriya's watchful eye, the NKVD undertook an investigation into Korolev's case, which concluded that he would be deprived of his freedoms for the next eight years. Although the verdict saved him from a return trip to the death camps, it was another cruel blow.

Tupolev had also been imprisoned during the purges, but because of the impending war Stalin took an interest in those who had worked or studied under him, and he ordered Tupolev to prepare a list of those who could be useful for work in the aeronautical industries. On that list was Korolev. He was transferred to an aviation design bureau located in

Stakhanov village near Moscow, where he was assigned to work under Tupolev. Korolev later said:

> *We were taken to the dining room, heads turned to our direction, sudden exclamations, people*
> *ran to us. There were so many well-known, friendly faces.*

Glushko, meanwhile, had been sentenced to a far less arduous eight years in a prison near Moscow – part of a larger network that held the scientific intelligentsia. The inmates called such places *sharashka*, meaning something sinister and based on lies.

Gagarin's War

The Second World War was not going well for Stalin. In October 1942 German artillery units shelled Klushino, some 100 miles (160 km) to the west of Moscow, and soon columns of troops passed through the village. Gunfire echoed in the surrounding woodlands as partisans confronted the advancing soldiers. The German attacks were met first with resistance and then a retreat as they advanced deeper and deeper into Russian territory. Two young brothers crept through the woods outside Klushino after the battle. One of them later recalled:

> *We saw a Russian colonel, badly wounded but still breathing having been lying where he fell for*
> *two days. German officers went to where he was, in a bush, and he pretended to be blind. Some*
> *high-ranking officers tried to ask him questions but he said he couldn't hear them very well and*
> *could they move closer. When they came closer and bent right over him he blew a grenade he'd*
> *hidden behind his back. No one survived.*

The Nazis terrorized the locals, subjecting them to summary execution; if ammunition was short, they used bayonets. The brothers lived dangerously, scattering broken glass on the road to burst the tyres of German trucks, and sometimes pouring dirt into their car batteries and petrol tanks. On one occasion a German offered one of the brothers some chocolate. Then he took him and strung him up in an apple tree to hang him. Their mother came running to confront the German, who brandished a rifle at her. Fortunately, he was called away at that point and the mother rushed to the tree, praying it was not too late. She was just in time to save her son.

The family had been evicted from their home and were living in a hole in the ground. The boy's limp body was dragged there while he slowly recovered. Frightened, cold and hungry they huddled in the dirt. There was Alexei the father, Anna the mother and four children. The eldest child was Valentin, who was 18, and the youngest was Boris, who was just six years old. It was Boris who had been saved from hanging. There was also a 15-year-old daughter called Zorya. Their other child was an eight-year-old boy called Yuri. The family's name was Gagarin. Valentin remembers how Yuri had changed as a result of the war; he became serious, introverted and deep:

> *Many of the traits of character that suited him in later years as a pilot and cosmonaut all*
> *developed during that time, during the war.*

'Many of the traits of character that suited him in later years as a pilot and cosmonaut all developed during that time, during the war.'

VALENTIN GAGARIN

Rockets for War

Meanwhile, at Peenemünde on Germany's Baltic coast, a rocket stood ready for launch. Slim and tall and with its tail adorned with fins, its metal sides became frosted as vapour danced around the chilled fuel tank. Suddenly a flame appeared at its base, and reddish-yellow smoke billowed in all directions. The retaining cables fell away as the rocket rose into the air. The flame became more intense and the rocket started to arc over the Baltic, effortlessly surpassing the speed of sound. After one minute, controllers on the ground sent a cut-off signal and engineers watched through binoculars as the flame died. The rocket was now over 20 miles (32 km) away. At this, Walter Dornberger wrote that his heart was beating wildly and that he wept with joy. Later he told the engineers that they had proved it would be possible to build piloted missiles or aircraft that could fly at supersonic speeds:

> Our rocket today reached a height of nearly 60 miles. We have invaded space and shown that
> rocket propulsion is practical for space travel.

Dornberger thought of the possibilities of space travel, but this was a time of war and the rocket was a weapon – a wonder weapon for the Third Reich. Initially the Germans called it the *A-4,* but it was renamed the *V-2.*

The British Air Force pounded Peenemünde in August 1943, so production of the *V-2* moved to an underground oil-storage depot in the Harz Mountains, near Nordhausen. The Nazis used slave labour working with pickaxes to enlarge the caverns. On 8 September 1944 two *V-2s* were launched from a site near the Hague in Holland, intended for a site about a mile (0.6 km) from Waterloo Station in London. It landed in Chiswick, killing two people. Von Braun told his staff:

> Let's not forget that this is the beginning of a new era, the era of rocket-powered flight. It seems
> that this is another demonstration of the sad fact that so often new developments get nowhere
> until they are first applied as weapons.

When they heard that the *V-2* had hit London those responsible drank champagne, with von Braun saying: 'Let's be honest about it. We were at war, although we weren't Nazis, we still had a fatherland to fight for.' He later commented on the rocket's performance: 'It behaved perfectly, but on the wrong planet.'

Race for the V-2

The West and the Soviets were both impressed with the *V-2*. As the war drew to a close it became a top priority to get their hands on the weapon, its technology and the engineers who built it. In a letter dated 13 July 1944 the British prime minister Winston Churchill requested Stalin's co-operation in locating and retrieving *V-2* components that the Germans were leaving behind in their retreat. Stalin ordered the formation of a secret group to collect any rocket remains.

At first they obtained some sparse but significant items, such as a combustion chamber and parts of propellant tanks. The pieces were sent back to Moscow where a group of engineers started to examine them. Among the group was Vasili Pavlovich Mishin, a specialist in control systems who, 20 years later, would lead the forlorn Soviet programme to land a cosmonaut on the Moon. The German rocket was far in advance of any technology possessed by the Soviets, or indeed anyone else. But they failed to recognize its full implications. They would eventually pay the price for thinking that long-range aircraft would be a superior weapon to a missile. However, some early investigations into rockets were being carried out in the missile development project led by a 30-year-old mathematician named Vladimir Nikolayevich Chelomei, who was later to play a vital role in the space programme.

In March 1945 the Pentagon sent a request to Colonel Holger Toftoy, Chief of Army Ordinance Technical Intelligence in Europe, for 100 operational *V-2*s. Toftoy sent Robert Staver to get the *V-2*'s blueprints and documents and to find its engineers.

Von Braun left the Harz complex just hours ahead of the Russians. 'We feared the Russians, despised the French and didn't think the British could afford us.' He planned to surrender to the Americans.

Stalin may have played a role in diverting troops towards Peenemünde rather than Berlin in the final months of the war. Just days after Hitler's suicide in Berlin, an infantry unit led by Major Anatole Vavilov from the Second Belorussian front took control of Peenemünde. The place was deserted and almost empty.

A German *V-2* rocket, the world's first ballistic missile, at Peenemünde.

1933
Soviet Union launches its first liquid-fuelled rocket

1934
December Wernher von Braun launches two *A-2* rockets from Borkum Island in the North Sea

1936
9 May Sergei Korolev oversees the first test of '216' winged missile

1944
Soviet troops capture remnants of a German *V-2* rocket

1950
26 April Sergei Korolev becomes chief designer of OKB-1, the Soviet long-range ballistic missile programme

1951
29 July The first launch of the Russian 'geophysical' rocket carrying live animals on board

1957
August First successful test flight of the Russian *R-7* ICBM

Stalin was furious, and was reported to have said:

This is absolutely intolerable. We defeated the Nazi armies, we occupied Berlin and Peenemünde: but the Americans got the rocket engineers. What could be more revolting and more inexcusable? How and why was this allowed to happen?

In June 1945 a group of Soviet engineers arrived at Peenemünde. Among them was a 33-year-old expert on guidance systems called Boris Chertok. He soon realized how far behind they had been in terms of technology. By the end of the war the most powerful operational Soviet rocket engine had a thrust of 1.7 tons, but the *V-2* had a thrust of 30 tons. One official knew the reason:

In Germany we realized that there were no arrests. As a result of repressions in the army and the scientific community our development had stopped at powder rockets.

The point was amplified when Soviet soldiers dug out from the rubble at Peenemünde a German edition of a book by Tsiolkovsky. On almost every page there were notes and comments made by von Braun. The Soviets also found in the archives of the Nazi Air Ministry drawings of a missile designed by Soviet engineers in the late 1930s.

Chertok and others arrived at Nordhausen to try to salvage whatever they could. But they needed more rocket experts to make sense of what they found. Glushko and Korolev were recommended, and soon they were also on their way to Germany.

Closing the Gap

At the end of the Second World War the Soviet Union may have had the most powerful land force in the world, but such forces suddenly became secondary following the bombing of Hiroshima and Nagasaki by atomic weapons. Just 18 days after Potsdam and 14 days after Hiroshima, on 20 August 1945, a secret decree of the Central Committee and the Council of Ministers called for the formation of the Special Committee on the Atomic Bomb to direct and co-ordinate all efforts on the rapid development of operational nuclear weapons. It was also necessary to have missiles to deliver them. Colonel General Mitrofan Nedelin and People's Commissar of Armaments Dimitri Ustinov were appointed by Stalin to lead the USSR's rocketry development. Nedelin, then aged 44, was a brilliant officer who had used solid-fuelled Katyusha rockets during the war. Korolev was placed in charge of developing long-range missiles. His first task was to build a Soviet copy of the *V-2* and then improve on it, but it was clear to him that creating a Soviet copy of the *V-2* would only serve as an interim measure. They needed better rockets of their own, even though the task of getting them approved by the government was going to be difficult.

In early 1945, Mikhail Tikhonravov, who in 1933 had worked with Korolev on the development of the first Soviet liquid-fuelled rocket, brought together a group of engineers to work on a design for a high-altitude rocket to carry two passengers to 120 miles (193 km). Called the VR-190 proposal, it was the very first established project in the Soviet Union aimed at launching humans into space. The plan envisioned the use of a modified *V-2* with a recoverable capsule for carrying two 'stratonauts'. Tikhonravov tried to obtain interest from the top:

> *Dear Comrade Stalin! We have developed a plan for a high-altitude Soviet rocket for lifting two humans and scientific apparatus to an altitude of 190 kilometres. The plan is based on using equipment from the captured* V-2 *missile, and allows for realization in the shortest time.*

Stalin was interested, at least initially, writing back: 'The proposal is interesting. Please examine for its realization.' But Tikhonravov's work stagnated. In 1947 it was renamed a 'rocket probe' and a year later a preliminary plan was presented for approval. Further work was allowed with one change – the launch of humans was dropped in favour of using dogs. The following year, the

> ## 'Dear Comrade Stalin! We have developed a plan for a high-altitude Soviet rocket ...'
>
> MIKHAIL TIKHONRAVOV

project was cancelled. Thus ended the first serious investigations in the Soviet Union of manned spaceflight. The issue would not re-emerge for several years.

Korolev took his argument for a space programme to Stalin himself. On 14 April 1947 he was escorted into the Kremlin to meet the Soviet leader in person for the first time. He later wrote of his frustration:

> *I had been given the assignment to report to Stalin about the development of the new rocket. He listened silently at first, hardly taking his pipe out of his mouth. Sometimes he interrupted me, asking terse questions. I can't recount all the details. I could not tell whether he approved of what I was saying or not. He said 'no' enough times that these 'no's became the law. But where rockets were being studied dreams of flight into space were not far behind.*

Plans for a Satellite Launcher

By early 1948 Tikhonravov was pushing forward with another idea – a satellite. He received little encouragement, but he was undeterred. That summer he read his report at the Academy of Artillery Sciences in the presence of a large group of prominent dignitaries from the military. Korolev was also there. The reaction of most was negative, but afterwards Korolev approached his old friend: 'We have some serious things to talk about.' Soon Korolev himself made plans to ask Stalin to fund the launch of an artificial satellite. The Soviets now had the *R-3* missile project, a rocket with a thrust of 120 tons designed to send a 3-ton warhead a distance of 1900 miles (3,000 km). Could it be the basis of a satellite launcher, he wondered?

In 1950 Tikhonravov tried once again to get official interest in the first detailed Soviet analysis of the requirements for launching an artificial satellite with a proposal entitled 'On the Possibility of Achieving First Cosmic Velocity and Creating an Artificial Satellite with the Aid of a Multistage Missile Using the Current Level of Technology'. It was presented at a special session of the Academy of Artillery Sciences. The reaction to this presentation was even worse than in 1948: some in the audience were hostile, others were sarcastic and many simply remained silent – Korolev was one of its few supporters.

Dogs in Space

In the summer of 1951 engineers led by Korolev converged on the isolated Kapustin Yar launch site in the Astrakhan Oblast for the first Soviet attempt at launching a living organism into space. Nine dogs were selected at first, from which Dezik and Tsygan were chosen.

The launch, using the new *R-4* missile, took place in the early morning so it would be illuminated by the sun during its ascent. The launch was successful and the dogs reached a maximum velocity of 2,600 miles per hour (4183 km per hour) and an altitude of 63 miles (101 km), officially entering space. The dogs also experienced four minutes of weightlessness.

After 188 seconds the payload section separated from the main booster and went into freefall until it reached an

The Russian *R-7* rocket, the first real intercontinental ballistic missile.

altitude of 3.7 miles (6 km), at which time the parachute deployed. Twenty minutes after lift-off the dogs were back on the ground barking and wagging their tails – the first living things recovered after a flight into space, and two months before the United States achieved a similar feat. Subsequent flights met with mixed results. Dezik and Lisa died when their parachute failed. After the second launch, it was decided that Tsygan, who had been Dezik's partner on the first flight, should not fly again. Instead, in early September, engineer Anatoli Blagonravov took her back to Moscow. Russia's first canine cosmonaut lived to a ripe old age, and Blagonravov and the dog would often be seen walking the streets of Moscow. In total, nine dogs were flown on six launches, three of them flying twice.

Stalin's death in March 1953 brought about the first change of leadership in the Soviet Union in more than 30 years, but the direction of the rocketry programme changed little. In early 1954 Premier Nikita Khrushchev instructed Minister Ustinov to dilute Korolev's monopoly in rocket design and construction. Ustinov came up with a plan to create two independent groups. Korolev's rival was to be the Experimental Design Bureau formed in the Ukraine and led by 43-year-old Mikhail Yangel.

The Rise of Russian ICBMs

Although rockets for space flight were important to the Soviet Union, what mattered more were intercontinental ballistic missiles (ICBMs). The first Soviet ballistic missile was the R-5. In February 1956, with a live atomic bomb in its nosecone, it was test launched from Kapustin Yar. Observers at the impact site of the 300-kiloton nuclear explosion in Kamchatka telephoned Korolev at the launch site informing him that: 'We have observed detonation.' The R-5 went into service and stayed in operation for 11 years. However, work was already under way on a more powerful ICBM, the R-7. At last the various factors needed to put a satellite into orbit were coming together. A suitable launcher was on the horizon, and Korolev's supportive colleague Marshal Nedelin had become Deputy Minister of Defence for Special Armaments and Reactive Technology. If a satellite were to lift off from Soviet soil, it would be Nedelin who would permit the use of a missile for such a project.

The R-7 was unlike anything created before. At the launch site, four conical strap-on boosters, each just over 19 metres (62 ft) in length, surrounded the central rocket core. It had a launch mass of 270 tons, of which about 247 tons was fuel. At lift-off, the total thrust was an impressive 398 tons. Korolev knew it could launch a satellite – he just needed permission from the authorities to allow him to do so.

'Man will conquer space soon'

THE FIRST SPACE TRAVELLERS

SPUTNIK, LAIKA AND EXPLORER 1
1957–1958

The Space Age truly began when the Soviet Union launched a satellite into orbit in 1957. This event amazed the world but also sent shock waves throughout the United States. If it was possible to send a satellite into orbit, how long before a warhead? Further successes by the Russians, and some failed rocket launches by the Americans, only increased these concerns.

Just outside Huntsville in Alabama the Americans had installed their greatest prize of the Second World War – the German rocket team. The secret removal of scientists from Nazi Germany – known as Operation Paperclip – was not only for the benefit of the Americans but also to deny the USSR. In April 1944 von Braun and his V-2 engineers had been ordered to Oberammergau in the Bavarian Alps under close SS guard, who had orders to shoot them rather than let them fall into enemy hands. However, von Braun convinced the SS that they should be dispersed into nearby villages so that the group would not be an easy target for bombers, and on 2 May he had fled. At the first opportunity his brother had approached an American soldier, calling out: 'My name is Magnus von Braun. My brother invented the V-2. We want to surrender.' The Americans were delighted. Von Braun was top of the 'black list' of German scientists and engineers they wanted to track down.

Working for the Americans

By the autumn, von Braun and his engineers were in the United States. Collating the V-2 documents and teaching the military what they knew about rockets, the team set about assembling and launching a number of V-2s from the White Sands Missile Base in New Mexico. Von Braun and his workers were not allowed to leave their quarters without a military escort, so he and his colleagues jokingly referred to themselves as 'Pops' – Prisoners of Peace. Clearly, von Braun's influence was less than he had hoped.

An official NASA photograph showing the historic first launch of a missile – a *Bumper V-2* – from Cape Canaveral, on 24 July 1950.

1945

Top German ballistic missile designers led by Wernher von Braun surrender to the US army in Germany. They would form the core of the missile development team at Redstone Arsenal in Huntsville, Alabama

1946

10 May The first *V-2* missile blasts off from White Sands, New Mexico

1956

September US army launches a *Jupiter-C* missile from Cape Canaveral, Florida

1957

1 May US navy conducts a test launch of the *Vanguard* rocket

4 October Russian *Sputnik*, the world's first artificial satellite, is launched

3 November *Sputnik 2*, carrying the dog Laika, is launched

1958

1 February US launches *Explorer 1* rocket into orbit

1 October The US Congress creates NASA

Between 1950 and 1956 von Braun and his team worked on the ICBM programme, which resulted in the *Redstone* rocket. He then developed the *Jupiter-C*, an improved *Redstone*. But he was frustrated. In a drawer in his desk he kept a notebook he had owned since he was 16 years old. Inside were sketches for a spaceship. But having demonstrated to the world his prowess at rocket technology it seemed that the US government was not interested in space. In the *Huntsville Times* of 14 May 1950 a headline reads: 'Dr von Braun Says Rocket Flights Possible To The Moon.' He also wrote articles for *Colliers Magazine* entitled 'Man Will Conquer Space Soon'.

He dreamed of 50 astronauts travelling to the Moon in three huge spacecraft and using the emptied holds of their craft as shelters. Astronauts would drive pressurized tractors hundreds of miles across the lunar surface, exploring its craters and plains. He imagined manned missions to Mars in which a fleet of ten spacecraft, each with a mass of almost 4000 tons, and some of them carrying a 200-ton winged lander, would descend on the Martian surface. To explain his vision he worked with Walt Disney on a series of films called 'Man in Space', which aired in 1955.

Von Braun knew that the *Jupiter-C* was capable of being modified to launch a satellite into orbit, but the US government had dictated that the navy should be first to launch a satellite – not the army that employed him.

Russia Joins the Race to Launch a Satellite

In the USSR the *R-7* needed a new launch site. The current one, Kapustin Yar, was too close to US radio monitoring sites in Turkey. The one chosen was at a place called Tyuratam in Kazakhstan. It was a remote, treeless, naked steppe, with a temperature of 45 °C (113 °F) in summer and sub-zero conditions in winter. The tsars had exiled undesirable citizens there. It had a curious connection with space; in the late 19th century the artisan Nikifor Nikitin was banished there for his 'seditious plans for a flight to the Moon'. Eventually the place was to be called Baikonur.

There was no space programme in 1954 when work on the *R-7* began in earnest, but that was to change as events unfolded swiftly. Armed with two large sketchbooks, Tikhonravov made an appointment to meet Georgi Pashkov, the missile department chief at the Ministry of Medium Machine Building. One of the books contained clippings from Western magazines, including von Braun's articles, with descriptions of American satellites. The other sketchbook contained detailed plans showing that the Soviets could achieve a satellite launch before the Americans, because the USSR had more powerful rockets. Pashkov was sufficiently impressed. The satellite study was approved.

'In my opinion, it will be possible to launch an artificial Earth satellite within the next two years.'

LEONID SEDOV

Competing with the Americans

But the Soviets were not the only ones planning to launch a satellite. In the spring of 1950, a group of American scientists led by James Van Allen met to discuss the possibility of an international scientific programme to study the upper atmosphere and outer space using rockets, balloons and ground observations. Soon the idea expanded into a worldwide programme timed to coincide with the anticipated intense solar activity from July to December 1957. They called it the International Geophysical Year (IGY). At a subsequent meeting in Rome in 1954, Soviet scientists silently witnessed the approval of an American plan to put a satellite into orbit during the IGY. In July the following year President Dwight D. Eisenhower's press secretary James C. Hagerty said that the United States would launch 'small Earth-circling satellites'.

Academician Leonid Sedov, Chairman of the Commission for the Promotion of Interplanetary Flights, USSR Academy of Sciences, called a press conference the same day at the Soviet embassy in Copenhagen at which he announced: 'In my opinion, it will be possible to launch an artificial Earth satellite within the next two years.' But a sceptical Soviet leadership needed to be convinced.

During a tour of Korolev's rocket factory by Premier Khrushchev, Korolev earnestly tried to explain to him the use to which his rockets could be put for research into the upper atmosphere. Somewhat out of his depth, the Soviet leader nevertheless expressed polite interest, although it was clear that most of the guests were becoming bored with the proceedings. Detecting that his guests were in a hurry to leave, Korolev quickly moved ahead and pointed everyone's attention to a model of an artificial satellite. Invoking the name of a legendary Soviet scientist, Korolev explained that it was possible to realize the dreams of Tsiolkovsky with the R-7 missile. Korolev pointed out that the United States had stepped up its satellite programme, but that compared with the 'skinny' American rocket, the Soviet R-7 could significantly outdo them. Khrushchev began to show more interest, and asked if such a plan might not harm the R-7 weapons research programme. Korolev said that all the Soviets needed to do was replace the warhead with a satellite. Khrushchev hesitated, but then said: 'If the main task doesn't suffer, do it.' Korolev had the green light at last. The USSR Council of Ministers issued a decree on 30 January 1956 calling for the creation of an artificial satellite, designated 'Object D', and approving its launch in 1957 in time for the IGY.

Early Russian Failures

As 1956 drew to a close, Korolev was exhausted by the constant travel from his factory at Kaliningrad, near Moscow, to Kapustin Yar and Baikonur. He was also worried that the Americans would beat him to his goal of launching a satellite. In September the US army had launched a *Jupiter-C* missile from Patrick Air Force

Base at Cape Canaveral, Florida. Had the missile been fitted with an additional rocket stage it would have been possible for it to launch a satellite. Korolev mistakenly believed that this had indeed been a secret attempt to put a satellite in space.

The results of static testing of the R-7 engines on the ground showed they were not as powerful as Korolev had hoped. But perhaps he was making things too difficult for himself, he wondered? Instead of launching a fairly large satellite of 1.5 tons, why not launch something simpler on the first orbital attempt? So he asked for permission to launch two small satellites, each with a mass of 40 to 50 kilograms (88 to 110 lbs), during the period of April to June 1957, the start of the IGY.

Fuelling for the R-7's first flight began on 15 May 1957, under the direction of Georgi Grechko, a 26-year-old engineer who would fly into space himself 18 years later. The most difficult part was handling the liquid oxygen, which was maintained at a temperature of –190 °C. The process took nearly five hours. When the time for the launch came the rocket lifted gracefully into the sky, but at T+98 seconds the rocket engines cut out. Engineers later discovered that the strap-on boosters had broken away from the central core.

The 50-year-old Korolev was not in good health; he had a bad sore throat and had to take regular penicillin shots. His letters to his wife at the time were full of doubt and frustration:

> When things are going badly, I have fewer 'friends'. My frame of mind is bad. I will not hide it. It
> is very difficult to get through our failures. There is a state of alarm and worry.

The second R-7 rocket was taken to the launch pad in early June after modifications. This time there were two launch aborts, traced to errors in the rocket's assembly. A third rocket was moved to the pad for launch on 12 July. This time it again lifted off into the sky, but at T+33 seconds all four strap-on units fell off. This was the lowest point for Korolev. There was talk of cancelling the entire programme, which would end his career. He wrote to his wife: 'Things are not going very well again. Things are very, very bad.'

A Question of National Pride

Wernher von Braun knew nothing of Korolev's existence, yet he had some notion of the technological struggles taking place behind the Iron Curtain. Following his successful launch of the *Jupiter-C,* he studied charts of its flight path. It had reached an altitude of 700 miles (1126 km). He knew that if it had been fitted with a fourth stage it could have put a satellite into orbit.

It was intensely frustrating for von Braun. He could put a satellite into orbit at any time, but the US government would not let him. They actually prevented him by sending observers to his rocket tests to make sure he did not sneak one into orbit when they were not looking! President Eisenhower and the Joint Chiefs of Staff did not want a German – especially an ex-Nazi – to launch the first American satellite; they wanted the navy to do it. But he knew that the navy's *Vanguard* rocket was inferior to the *Jupiter-C,* and it was behind schedule. More than once he said that the navy would lose the race to the Russians because their rocket would not work. He even said they could paint '*Vanguard*' on the side of his rocket. But no one was listening to him.

Sputnik **Orbits the Earth**

In Kazakhstan another *R-7* was brought to the pad. It successfully lifted off on 21 August and this time all systems worked, the missile and its payload flying 4000 miles (6437 km) before the warhead entered the atmosphere over the target point at Kamchatka. Excited, Korolev stayed awake until three in the morning talking about the great possibilities that had opened up. Korolev, Glushko and the other chief designers had informally planned the satellite launch for the 100th anniversary of Tsiolkovsky's birth on 17 September, but that date was now unrealistic. The next *R-7* booster, this time with the *Sputnik* satellite onboard, was wheeled to the launch pad in the early morning of 3 October, escorted on foot by Korolev. He told his engineers:

Sputnik, the first satellite to orbit the Earth, being worked on by a Soviet engineer in the autumn of 1957 prior to its launch.

> *Nobody will hurry us. If you have even the tiniest doubt, we will stop the testing and make the corrections on the satellite. There is still time.*

At precisely 22 hours, 28 minutes and 34 seconds Moscow Time, the engines ignited. There were problems, but they were not major ones. Satellite separation from the core stage occurred at T+324.5 seconds and the first manmade object entered orbit around the Earth. The Space Age had begun. Korolev waited next to the communications van along with a huge crowd. There was cheering once the Kamchatka station picked up signals from *Sputnik*, but Korolev advised them to hold back their celebrations until it had completed one revolution of the Earth. Eventually the 'beep-beep-beep' of *Sputnik* was heard as it started its second orbit. State Commission Chairman Ryabikov waited until the second orbit was complete before telephoning Premier Khrushchev, who was visiting Kiev.

Later Korolev and a small group took off from Baikonur for Moscow. Most were exhausted and slept throughout the flight. After take-off the pilot of the airplane, Tolya Yesenin, came over to say to Korolev that 'the whole world was abuzz' with the launch. Korolev was invited into the pilot's cabin. When he returned he said: 'Comrades, you can't imagine – the whole world is talking about our satellite. It seems that we have caused quite a stir.'

In the morning edition of *Pravda* the news was exceptionally low key and was not the headline of the day. The Soviet media did not ascribe a specific name for the satellite, generally referring to it as *Sputnik*, the Russian word for 'satellite', loosely translated as 'fellow traveller'.

An American Failure

That same day the Society of Experimental Test Pilots was holding a symposium in the Beverly Hilton Hotel, California. Neil Armstrong, a young test pilot, was taking part. He was trying unsuccessfully to get the Los Angeles press interested in the various technical advancements in the test-flight world. Then he heard about *Sputnik*. Instantly he knew it would change the world. He watched on television as President Eisenhower seemed to completely miss the point by saying: 'What's the worry? It's just one small ball.' Perhaps, Armstrong thought, that was a façade behind which the president was hiding substantial concerns, because if the Russians could put an object into orbit, they could also put a nuclear weapon anywhere in the United States. For Armstrong it was humiliating that a country deemed to represent an evil regime in the eyes of the American people was overtaking them in technology, an area in which they themselves assumed they were leaders.

Von Braun had arranged a dinner party for the defence secretary designate Neil McElroy in the officers' mess at the Redstone Arsenal. After dinner the base public relations officer ran into the room and handed von Braun a piece of paper informing him that the Russians had launched a satellite. The event had even been picked up by an amateur radio ham in Huntsville. Von Braun was angry. He turned to McElroy:

> *We knew they were going to do it.* Vanguard *will never make it. We have the hardware on the shelf. We can put up a satellite in 60 days.*

'What's the worry? It's just one small ball.'

PRESIDENT EISENHOWER

He was right. On 6 December the navy launched their *Vanguard* rocket from Cape Canaveral. It reached an altitude of 1.2 metres (4 ft), fell and exploded. The small, 1.3-kilogram (2.9-lb) satellite was thrown clear, bleeping pathetically as it rolled away. The press called it *Kaputnik*.

America was shocked, however, although Eisenhower still seemed unconcerned. One member of the public summed it up when she was interviewed for television. Hesitantly, she said: 'We fear this.'

The First Earthling to Orbit Earth

Khruschev was surprised and delighted with the worldwide reaction to *Sputnik*. He asked Korolev what else he could do. Korolev knew the answer. 'We can launch a dog,' he replied.

The first living creature to orbit Earth was found on the streets of Moscow, scavenging for food in dustbins. She was a small mongrel dog about three years old, malnourished, but with a good temperament. She was taken in by scientist Oleg Gazenko, who called her Kudryavka, or Little Curly Haired One. Originally, ten dogs were in the running to be chosen for the flight, all of them trained at the air force's Institute of Aviation Medicine for rocket flights into the upper atmosphere. They were subjected to many tests and procedures: exposed to noise and vibration, swung around in centrifuges, kept in progressively smaller cages for up to 20 days and trained to eat high-nutrition gel. They were finally reduced to one, Kudryavka, now known as Laika, which means 'barker'. Air force doctor Vladimir Yazdovskiy recalls:

> *Laika was a wonderful dog, quiet and very placid. I once brought her home and showed her to the children. They played with her. I wanted to do something nice for the dog. She had only a very short time to live.*

Konstantin Feoktistov, a promising 32-year-old engineer later to become a cosmonaut himself and to play an important role in the design of the *Salyut* and *Mir* space stations, was placed in charge of the engineering details for the mission. As a 16-year-old he acted as a scout for Soviet partisans in his home town of Voronezh in southwestern Russia. He had been captured, shot and left for dead. But the bullet had only grazed his throat, and he crawled out of a pit of corpses to reach safety under cover of darkness.

The dog Laika, the first animal in space, inside a mockup of the Soviet capsule *Sputnik 2*.

Sputnik 2 was a rushed job, assembled in less than a month. All who worked on it knew it was unsatisfactory, being designed to fulfil the needs of propaganda instead of science. It had a crude life-support system, food for seven days in the usual gelatinous form and a bag to collect waste. It was so cramped there would be no room for Laika to turn around. Neither was there a re-entry mechanism. It was to be a one-way trip.

Laika was placed in the capsule three days before launch. Engineers monitored her to ensure she did not become too distressed. The lift-off, on 3 November, did not go well. Although *Sputnik 2* was placed into orbit, some of its thermal insulation tore loose and the temperature inside rose to above 40 °C (104 °F). Telemetry indicated that Laika was overheating and agitated. But after five hours all went quiet. Scientists had planned to euthanase her with poisoned food. For years afterwards it was reported that she died when her oxygen ran out, but in reality she succumbed very rapidly to heat stroke – an unpleasant death. She was cremated on 14 April 1958 when *Sputnik 2* burnt up during re-entry. Earth's first space traveller had completed 2570 orbits. In 1998 Oleg Gazenko expressed regret at the mission:

> The more time passes, the more I am sorry about it. We shouldn't have done it. We did not learn enough from the mission to justify the death of the dog.

But Korolev knew that they would not stop at dogs. After winning the race to put the first artificial satellite into space they now had plans to launch the first spaceman. Korolev was determined that they would beat the Americans, who he felt sure would react to the *Sputnik* shock by trying to overtake them. In his mind he had a goal of December 1960 as the earliest they could put a human into orbit. Clearly they were ahead of the US, but would it stay that way?

The Space Race Has Begun

As a result of the *Kaputnik* fiasco von Braun got the go-ahead to launch a small satellite, and on 1 February 1958 *Explorer 1* was successfully carried into orbit by a modified *Jupiter-C* rocket. 'How, I wonder, would the USSR have reacted if the United States had beaten them into orbit – which they could have so easily done?' von Braun is reported to have said. He knew that the United States should have been the first nation to place a satellite into orbit. Korolev knew it, too. But now the space race had begun. Who would be the first to put a man in space?

'I can see the clouds, everything. It's beautiful'

ASTRONAUTS RACE COSMONAUTS

THE MERCURY 7 AND YURI GAGARIN
1959–1961

Stunned by the early successes of the Russian space orbits, the United States urgently set up a task group with the express purpose of being the first nation to put a human in space. But the group of astronauts chosen for this mission – known as the Mercury 7 – were to find themselves thwarted when Russian cosmonaut Yuri Gagarin also claimed this prize for the Soviets, beating the American attempt by a mere three weeks.

The United States realized that a manned spaceflight was the next big step. Coming round at last to the importance of spaceflight, President Dwight D. Eisenhower had assigned the National Advisory Committee for Aeronautics (NACA) to develop and carry out manned spaceflights. Two months later, in July 1958, the NACA became the National Aeronautics and Space Administration (NASA). Within a week its director T. Keith Glennan approved plans for a manned launch, giving the responsibility to the Space Task Group based at the Langley Research Center in Virginia, headed by Robert Gilruth. There were already preliminary sketches of a small capsule that was jokingly said to have been designed to be worn, not flown. The initiative, later named Project Mercury, began on 7 October 1958. What was needed next were astronauts.

Mercury 7 Revealed to the Public

One of those selected, Alan Shepard was destined to narrowly miss becoming the first person in space.

> Shepard: *April 9 1959 was one of the happiest days of my life. That was the day on which we all congregated officially as the US first astronaut group. We had been through a selection process, obviously, previous to that time. But that was the day we first showed up officially as the first astronauts in the United States.*

They were called the Mercury 7, and they were: Scott Carpenter, Gordon Cooper, John Glenn, Virgil Grissom, Wally Schirra, Alan Shepard and Deke Slayton.

The *Mercury 7* astronauts: left to right, back row, Alan Shepard, Virgil 'Gus' Grissom and L. Gordon Cooper; front row, Walter Schirra, Donald 'Deke' Slayton, John Glenn and Scott Carpenter.

Shepard: *Glenn, of course, I had known before; Schirra I had known before because of our navy connections. So I knew there was a lot of talent there, and I knew that it was going to be a tough fight to win the prize. Well it was an interesting situation because, as I say, I was friendly with several of them. And on the other hand, realizing that I was now competing with these guys, so there was always a sense of caution I suppose – particularly talking about technical things. Now in the bar everything changed, but in talking about technical things there was always a sense of maybe a little bit of reservation, not being totally frank with each other, because there was this very strong sense of competition. There were seven guys competing for the first job, whatever that turned out to be. Seven guys going for that one job. So on the one hand there was a sense of friendliness and maybe some support but on the other hand, 'Hey, I hope the rest of you guys are happy because I'm going to make the first flight.' I suspect my thoughts generally reflected those of the other chaps.*

John Glenn, who would become the first American to orbit the Earth:

They made every measurement you can possibly make on the human body, all the usual things you'd think about, plus all the other things that would occur in any natural physical exam, and then things like, oh, cold water in your ear. You sit, and you have a syringe, and you put cold water in your ear for a period of time. This starts the fluids in your inner ear, in the semicircular canal, starts them circulating because of the temperature differential, starts them circulating, and so you get the same effect as though you'd been spun up on a chair or something like that until you are extremely dizzy, and you had nystagmus, as it's called, your eyes want to drift off. You can't keep them focused on a spot. And then they would measure how long it took for us to recover from that. There was supposed to be some correlation to something, whatever it was. They had a lot of tests like that.

The Mercury 7 were paraded before the press and heralded by the American public as heroes. Viewed from today's perspective the press conference was a strange event. John Glenn did most of the talking, while Alan Shepard was perhaps the wittiest. Several of them smoked during the interviews and all, when asked, gave their home addresses, clearly something that would not happen today. A few weeks after being chosen they moved to Langley Field, Virginia, where they were shown the prototype of their spacecraft:

Shepard: *It didn't look very much like an airplane, but if you were going to put a pilot in it, it was going to have to fly somehow like an airplane, and when you have a strange new machine, then you go to the test pilots. That's what they were trained to do, and that what's they had been doing.*

At first the capsule did not have windows; engineers had thought them unnecessary since they would compromise its structural strength. The astronauts would have none of it. They insisted on windows and on something to do. Like their Soviet counterparts, the designers wanted almost everything to be automatic, with the astronaut acting as a passenger and doing little except in an emergency. The future space travellers were scathing, calling this method of flying 'chimp mode'.

No one knew how the human body would react under conditions of zero gravity (zero G). Some thought there would be only small effects, while others took the view that in zero G an astronaut might not be able to breathe or swallow properly and could become hopelessly disorientated.

> *'... believe me, it's a lot harder to land a jet on an aircraft carrier than it is to land a Lunar Module on the Moon. That's a piece of cake, that Moon deal!'*
>
> ALAN SHEPARD

Shepard: *This is a generalization, but it's something which I'd been doing for many, many years as a navy pilot, as a carrier pilot; and believe me, it's a lot harder to land a jet on an aircraft carrier than it is to land a Lunar Module on the Moon. That's a piece of cake, that Moon deal! And here you had, yes, a new environment, but you know, for fighter pilots who fly upside-down a lot of the time, zero gravity wasn't that big a deal.*

The First Astronaut is Chosen

Having lost the first lap in the space race to the Soviet Union, many in the United States still believed that they could win the race to put a person in space, but first there was the difficult decision to be made – who would that person be? All of the Mercury 7 were in with a chance. Shepard remembers the day the choice was made:

We had been in training for about 20 months or so, toward the end of 1960, early 1961, when we all intuitively felt that Bob Gilruth had to make a decision as to who was going to make the first flight. And, when we received word that Bob wanted to see us at 5.00 in the afternoon one day in our office, we sort of felt that perhaps he had decided. There were seven of us then in one office. We had seven desks around in the hangar at Langley Field. Bob walked in, closed the door, and was very matter-of-fact as he said: 'Well, you know we've got to decide who's going to make the first flight, and I don't want to pinpoint publicly at this stage one individual. Within the organization I want everyone to know that we will designate the first flight and the second flight and the backup pilot, but beyond that we won't make any public decisions. So, Shepard gets the first flight, Grissom gets the second flight, and Glenn is the backup for both of these two sub-orbital missions. Any questions?' Absolute silence. He said: 'Thank you very much. Good luck,' turned around, and left the room. Well, there I am looking at six faces looking at me and feeling, of course, totally elated that I had won the competition. But yet almost immediately afterwards

1959

9 April NASA announces the selection of America's first seven astronauts for the Mercury programme

1960

15 May First Vostok mission fails to return to Earth as planned

28 July Launch of second Vostok mission fails, killing dogs on board

19 August Successful Vostok mission launched with dogs Belka and Strelka on board

24 October Russian *R-19* rocket explodes on launch pad

1 December Russian space dogs Pchelka and Mushka launched into orbit

22 December Russian dogs Kometa and Shutka survive a launch failure

feeling sorry for my buddies, because there they were. I mean, they were trying just as hard as I was and it was a very poignant moment because they all came over, shook my hand, and pretty soon I was the only guy left in the room.

Selecting Russian Cosmonauts

The publicity and high expectations aroused by the Mercury 7 had not gone unnoticed in the USSR. Worried that they would not maintain their apparent lead, Khrushchev called a meeting of all the key people in the Soviet space effort, saying:

Your affairs are not well. You should quickly aim for space. There are strong levels of work in the USA and they'll be able to outstrip us.

Soon, representatives of the military and the Academy of Sciences met to talk about selecting the first cosmonaut. Where should they look for candidates? The air force, the navy and even racing drivers were briefly considered, but the air force insisted they had to be pilots. So the search began.

Only men were to be considered. They had to be between 25 and 30 years of age, no taller than 1.70 to 1.75 metres (5.6 to 5.7 ft), and with a weight of no more than 70 to 72 kilograms (154 to159 lbs) so that they could fit into the small capsule being designed. Two air force doctors were appointed to run the selection process, and teams were sent to air force bases in the western Soviet Union to look for candidates. Those who passed the initial selection were interviewed, but none was aware of the true nature of the mission, which was described as 'special flights'. Just over 200 passed this early screening, and they were then sent in groups of 20 for further testing at the Central Scientific Research Aviation Hospital in Moscow.

Testing under the 'Theme No. 6' programme involved spinning the pilot in a stationary seat to test the vestibular system (a sensory mechanism involved in balance and spatial orientation) and subjecting him to low pressure and increased gravity in a centrifuge. At the end of 1959 the number of candidates had been narrowed down to 20, and these were sent back to their units to await further orders. Of the group, five were not between the ages of 25 to 30, but this condition was waived because of their strong performances. In the end, none of those selected was a test pilot. One of them, Vladimir Komarov, had some experience as a test engineer flying new aircraft, but the most experienced pilot among them, Pavel Belyayev, had accrued only 900 hours of flying time. Others, such as Yuri Gagarin, had flown only 230 hours.

Twelve of the 20 cosmonaut candidates undertook final medical tests at the Central Scientific Research Aviation Hospital. Later, Gagarin recalled having seven eye tests as well as a series of mathematical tests during which a voice whispered into his headphones giving him the wrong answers. His heart was the focus of the tests. 'We were tested from top to toe,' he said. The training was divided between academic disciplines and physical fitness. They attended classes covering rockets, navigation, radio communications, geophysics and astronomy. Within a few weeks each candidate had made 40 to 50 parachute jumps. One of the USSR's top test pilots, Mark Gallay, supervised their aircraft training. Under his direction they flew parabolic trajectories to simulate weightlessness for periods up to 30 seconds in specially equipped aircraft. Soon they all moved about 25 miles (40 km) northeast to a new suburb of Moscow, which was renamed Zelenyy, meaning 'Green'. Today it is better known by a more recent designation – Zvezdny Gorodok, or 'Star City'.

Crude spacecraft simulators were installed at the new training base, and because it was believed to be inefficient to train all 20 on one simulator, a choice needed to be made. A group of six would undergo accelerated training, and from them the selection of the first cosmonaut would be made. One of them – Gagarin, Kartashov, Nikolayev, Popovich, Titov or Varlamov – would be the first person in space, or so they hoped. Korolev visited the centre for the first time in June 1960. The cosmonauts had only learned of his existence a few months earlier, and even then he was only called the 'chief designer'. He carried with him diagrams of the space capsule. Korolev had settled on a simple yet remarkable design concept for the first spacecraft: it was to be spherical, with one side heavier than the other. The heavier side would be fitted with a heat shield. Its extra weight meant that this side would automatically turn and face forward on re-entry, shielding the craft from the searing heat of friction as it entered the Earth's atmosphere. Soon it was named Vostok or 'East'.

Testing in Earnest

The US watched the Russian activities as closely as possible. Radio listening stations in Turkey were alerted that something was going on. In April a U-2 spyplane of the CIA (Central Intelligence Agency) took off from Peshawar Airport and flew over the Semipalatinsk test site, the air defence forces near Saryshagan, and then the Baikonur complex.

In the USSR there was considerable pressure to accelerate the schedule, primarily because of the stream of news about Project Mercury. By the early summer of 1960, NASA officials were expecting to fly the first suborbital piloted Mercury craft early in 1961. Korolev was determined that the first piloted Vostok craft would be in orbit before the Americans' first launch. The deadline was specified in an official document from the Soviet government dated 4 June 1960 and entitled 'On a Plan for the Mastery of Cosmic Space'; all testing for a piloted Vostok flight was to be completed by December 1960.

In early summer the Vostok capsule was transported to Baikonur for a test launch. It was not fitted with a heat shield or an ejection seat, as the aim was only to test its basic elements – in particular the complex but essential Chayka orientation system that would orientate the spacecraft correctly for re-entry. Although this test capsule would burn up on re-entry, telemetry data would indicate whether or not it had been

'There are sea ships, river ships, air ships, and now there'll be space ships!'

SERGEI KOROLEV

properly aligned before it happened. As soon as it reached orbit, the State Commission issued a communiqué for the Soviet press. But first there was an issue over what to call the vehicle. Korolev said: 'There are sea ships, river ships, air ships, and now there'll be space ships!' Although the term 'space ship' was used in the official TASS news agency report for the first time, there was no indication that the mission had any relevance to a manned spaceflight. But officials in the US knew what it implied. The unmanned Vostok test versions were launched from Baikonur into a suborbital trajectory to splash down in the Pacific Ocean. As the Soviet navy plucked the capsules out of the sea, the US navy watched from close by.

Safely Back Home

But all did not go well with the first non-manned versions of the capsule. On re-entry, the most critical part of the flight, the retrorocket fired on time but because the spacecraft was pointing in the wrong direction it went into a higher orbit, where it stayed for more than five years before coming back to Earth. The problem was tracked down to a faulty sensor, which was removed from the version of the Vostok that was to carry a cosmonaut.

Accordingly, a second test Vostok was prepared, this time with two dogs, Chayka and Lisichka, on board. They were launched in July 1960, but the mission immediately ran into serious problems. Some 20 seconds after launch the rocket began to veer sideways. One of the strap-on engines exploded. The emergency escape rockets on top of the capsule were fired to get the 'crew' up into the air and away from the launch pad as quickly as possible, but it was already too late; the dogs were dead.

Undaunted, Korolev ordered the next Vostok test mission to carry two more dogs, Belka and Strelka, along with a biological cargo including mice, rats, insects, plants, fungi, cultures, seeds of corn, wheat, peas, onions, microbes, strips of human skin and other specimens. Lifting off on 19 August it reached orbit, but the television pictures relayed back were poor. At first, the dogs appeared still, later they became more animated, but their movements were odd and they were clearly ill. Belka squirmed and vomited. Was it possible that living things could not stand more than a single orbit in space? They parachuted into the Orsk region in Kazakhstan after a one-day, two-hour spaceflight, making Belka and Strelka the first living beings to be recovered from orbit. The spacecraft itself was only the second object retrieved from orbit, the American *Discoverer 13* having pre-empted it by nine days.

It was recommended that one or two further Vostok test flights be carried out in October–November 1960, followed by two automated missions of the Vostok flight configuration in November–December 1960. Korolev's plan was that by December the cosmonauts would be ready for a manned flight later that month, in time to beat a Mercury launch. Then disaster struck.

Disaster on the Launch Pad

In his Moscow office, Korolev received a late-night call from the Baikonur complex on a secure telephone line informing him that there had been a major accident, the catastrophic nature of which only became clear as more information arrived throughout the night. It involved a rocket designed not by Korolev, but by his rival, Mikhail Yangel.

Yangel's group was competing with Korolev to build a new generation of ballistic missiles, so he had brought his first missile, the *R-16*, to Baikonur in mid-October for its maiden launch. After the relative failure of Korolev's *R-7* rocket (now being modified to carry the first cosmonaut) as an operational ICBM, there was a lot of pressure to bring the technically superior *R-16* to operational status. It would finally justify Premier Khrushchev's bluster and bragging about Soviet rocket might. Just days before the planned launch, in a speech at the United Nations, he boasted that strategic missiles were being produced in the USSR 'like sausages from a machine', even though this was not true. Many important officials were at Baikonur to witness the first *R-16* launch, among them Strategic Missile Forces Commander-in-Chief Nedelin, who chaired the State Commission for the *R-16*.

There had been problems with the highly toxic propellants prior to launch, the worst of which involved fuelling procedures. These not only caused great consternation but also resulted in a whole day being lost due to a leak being discovered. On the orders of the State Commission, all repairs to the missile were to be carried out when it was fully fuelled – a very dangerous situation. After they were completed, and just 30 minutes before the launch on 24 October, there were approximately 200 officers, engineers and soldiers near the launch pad, including Marshal Nedelin, who scoffed at suggestions that he leave the area, saying: 'What's there to be afraid of? Am I not an officer?' Yangel had gone into a bunker to smoke a last cigarette before launch. It saved his life.

An inquiry later determined that the second stage rockets of the *R-16* ignited due to a control system failure. The flames cut into the first-stage fuel tanks beneath, which then exploded. Automatically activated cinema cameras filmed the explosion. People near the rocket were instantly incinerated, while those further away were burned to death or were poisoned by the toxic gases. When the engines fired, most of the personnel ran to the perimeter but were trapped by the security fence and then engulfed in the fireball of burning fuel. Deputy Chairman of the State Committee of Defence Technology, Lev Grishin, who had been standing next to Nedelin, ran across the molten tarmac and jumped onto a ramp from a height of 3.5 metres (11.5 ft), breaking both legs in the process and dying later of burns. As usual the incident was kept secret and Marshal Nedelin was said to have died in an aircraft accident, a deception the Soviets maintained until early 1989. About 130 people perished as a result of the explosion, many of whom were identified only by medals on their jackets or rings on their fingers.

Vostok Fails

Keen to forget the accident, the authorities granted Korolev permission to launch the fourth and fifth in the Vostok test spacecraft series. The first was launched without incident on 1 December 1960, into an orbit exactly mimicking the one planned at the time for a manned mission. Aboard were two dogs, Pchelka and Mushka. After about 24 hours the main engine fired to begin re-entry, but it fired for a shorter period than planned and the indications were that the landing would overshoot Soviet territory. To prevent it falling into the wrong hands a self-destruct system operated, blowing up the spacecraft and its passengers. Soon the fifth Vostok test spacecraft, carrying the dogs Kometa and Shutka, was sent on its way, but the third-stage engine prematurely cut off at 425 seconds. The emergency escape system went into operation, and the payload reached an altitude of 133 miles (214 km) and landed about 2175 miles (3500 km) downrange in the region of the Podkamennaya Tunguska river, close to the impact point of the 1908 Tunguska meteorite. There had been two consecutive failures of Vostok test flights and it was not possible to launch a cosmonaut by February 1961. Would the Americans, with their Mercury capsule, win after all?

Preparation for the First Manned Flight

In January 1961 the commission recommended the following order for flights: Gagarin, Titov, Nelyubov, Nikolayev, Bykovsky, Popovich. Gagarin was the favourite. One engineer said of him: 'He would never try to ingratiate himself, nor was he ever insolent. He was born with an innate sense of tact.' Earlier the Medical Commission had described his personality:

> *Modest, embarrasses when his humour gets a little too racy: high degree of intellectual development; fantastic memory; distinguishes himself from his colleagues by his sharp and far-ranging sense of attention to his surroundings: a well-developed imagination: quick reactions: persevering, prepares himself painstakingly for his activities and training exercises, handles celestial mechanics and mathematical formulae with ease as well as excels in higher mathematics: does not feel constrained when he has to defend his point of view if he considers himself right: appears that he understands life better than a lot of his friends?*

Indeed, when the cosmonauts carried out an informal survey to choose who they would like to fly first, all but three named Gagarin. Also in the running for the first flight was 25-year-old Gherman Titov, who had served as a pilot in the Leningrad region. He struck many as being the most well-read of the finalists, liable to quote Pushkin or refer to Prokofiev. The last of the top three, 26-year-old Grigori Nelyubov, was perhaps the most talented and qualified of the group. He had influential supporters but was extremely outspoken. The cosmonauts were working hard, spending long periods away from home, and when they were home they could not talk about what they did. Gagarin's wife Valya later said that if she ever asked him what he was training for he would dismiss it with a joke. Once, when Yuri brought some of his cosmonaut friends home, she heard them saying that soon it would either be Yuri or Gherman.

In January and February 1961, preparations for the launches of the remaining two Vostok test missions progressed. Each was identical to the actual piloted Vostok, but would carry a single dog into orbit and a life-sized mannequin would be strapped into the main ejection seat. The mission was to last a single orbit, the same as planned for the first human flight. Khrushchev announced on 14 March during an interview:

> *The time is not far off when the first space ship with a man on board will soar into space.*

The first human-rated Vostok spacecraft lifted off successfully just days later, carrying the dog Chernushka together with mice, guinea pigs, reptiles, seeds, blood samples, cancer cells and bacteria. The ejection seat was taken up by a life-sized mannequin (called Ivan Ivanovich) dressed in a Sokol spacesuit. The main purpose was to test radio communications with the capsule; however, they knew the Americans would be listening. So it was decided to tape a popular Russian choir, then, if the Americans heard it they would not know it was a recording and not think it was a real human voice. The successful mission had lasted only one hour and 46 minutes. The way was now open for a manned flight.

The six key cosmonauts flew to the Baikonur complex on 17 March to witness the pre-launch operations of the final test, which went well. All re-entry procedures were conducted without any problems. Shortly afterwards a press conference was held in Moscow that relayed little information except that a successful test flight had occurred. In the audience were foreign journalists as well as, in the front row, Gagarin, Titov and the other cosmonauts, but of course none of the press knew that one of them would fly in space in just a few days.

Knowing that the time was near, Korolev, in a touching move, invited some of the original GIRD veterans to his offices just a month before the first manned launch. Many of them had not seen him for many years. Over vodka they spoke of the old times and their dreams of space travel. His guests knew nothing of his secret work, but when they had reminisced a while he ushered them into a nearby workshop. There, in the corner, was the polished silver cockpit of the Vostok spacecraft. As they beheld it, knowing that the age of manned spaceflight was about to dawn, some of them wept for joy.

Who Would be the First Cosmonaut?

Things were moving quickly. Leaving for Baikonur for the last time before the manned flight, the cosmonauts were ordered to tell their spouses that the launch was set for 14 April, three days later than actually intended, so they would not worry as much. But who was to be the first spaceman – cosmonaut number one?

The State Commission had addressed the question at a meeting on 8 April. Both Gagarin and Titov had performed without fault, with Gagarin ahead in the January examinations. Nikolai Kamanin, the air force's representative in the space programme, wrote in his journal:

> Both are excellent candidates, but in the last few days I hear more and more people speak out in favour of Titov and my personal confidence in him is growing, too. The only thing that keeps me from picking Titov is the need to have the stronger person for a second one-day flight.

Photographs and details of Gagarin and Titov were sent to the Central Committee. Khrushchev replied: 'Both are excellent. Let them decide for themselves.' Finally Kamanin, perhaps with the more arduous second flight in mind, nominated Gagarin as the primary pilot and Titov as his backup for the first flight, whose launch date was set as 11 or 12 April. Later Kamanin invited Gagarin and Titov to his office and told them that Gagarin was going to fly and that Titov would serve as his backup. Years later, when asked how he felt, Titov said it had been unpleasant. He had wanted to be the first, but somehow could see why

1961

6 January The first six Russian cosmonauts are selected

31 January Chimpanzee Ham is the first primate in space aboard US *Mercury-Redstone 2*

March Two test missions of human-rated Vostok spacecraft

12 April Russian cosmonaut Yuri Gagarin becomes the first man into space aboard *Vostok I*, experiencing the weightlessness of space for 108 minutes

they made the choice they had: 'Yuri turned out to be the person that everyone loved. Me, they couldn't love. I'm not lovable.'

Launch Day Approaches

In the early morning hours of 11 April the huge doors of the main assembly building at Baikonur rolled open to reveal the *R-7* booster positioned on its side on a converted railway carriage. Very slowly it moved out into the dawn air. Alongside, and watching every movement, strode a nervous Sergei Korolev. He walked around the rocket many times on its 2.5-mile (4-km) journey to the launch pad. This was his 'child'. This was history. In a couple of hours it reached the pad and was levered into an upright position with an access gantry positioned alongside. Korolev would not leave it until it left the Earth. Early in the afternoon Gagarin and Titov arrived for a last-minute rehearsal. Korolev was on the point of nervous and physical exhaustion. More than once that afternoon he had to be helped to a chair for a rest. Meanwhile, an army general at a base on the outskirts of Saratov received a phone call from the Kremlin telling him to organize the recovery of the world's first spaceman, who would be landing in his region tomorrow.

The night before launch Gagarin and Titov were assigned to a cottage near the pad area, which had previously been used by Marshal Nedelin. After a light meal they were in bed by 7.30 pm. Korolev checked on them periodically, being unable to sleep himself. Medical sensors were attached to both cosmonauts to monitor their vital systems, and strain gauges were attached to their mattresses to see how well they slept. The official history states that they both had a good night, but that is not true. Gagarin later said that he did not sleep a wink but worked hard to stay perfectly still lest the strain gauges on the bed indicated that he was restless and the mission be given over to Titov. Evidently Titov did the same. It was hardly the best preparation for the first manned spaceflight.

So many things were going through Korolev's restless mind; so many failure modes. Among the many worries perhaps the most troubling was the prospect of the rocket's third stage failing during the ascent to orbit, depositing the Vostok spacecraft in the ocean near Cape Horn on the southern tip of Africa, an area infamous for its constant storms. Korolev had insisted that there be a telemetry system in the launch bunker to confirm that the third stage had worked as planned. If the engine worked correctly, the telemetry would print out a series of 'fives' on tape; but if it had failed, there would be a series of 'twos'.

Preparation for Launch

Pre-launch pad operations began in the early hours of 12 April. By dawn officials and controllers had taken up their positions. Gagarin and Titov were woken at 5.30 am to be presented with a bunch of early wild flowers, a gift from the woman who had previously owned the cottage. After a short breakfast of meat

Yuri Gagarin on his way to the launch pad on the morning of 12 April 1961.
Seated behind him is backup pilot Gherman Titov.

paste, marmalade and coffee, doctors examined the cosmonauts, and assistants helped Gagarin and Titov into their cumbersome Sokol spacesuits followed by a bright orange coverall. Titov was dressed first since they did not want Gagarin to overheat. Soon they were on the bus to the launch site accompanied by 11 others including cosmonauts Nelyubov and Nikolayev and two cameramen. The film shows Gagarin taking his seat behind a small table. Titov walks past and sits behind Gagarin; hardly anyone seems to notice him.

At the pad, Gagarin and Titov were greeted by Korolev, Kamanin and other officials. Korolev, having not slept at all, looked fatigued as he watched Gagarin. After the embraces Gagarin went to the service elevator, where he halted and waved before the two-minute ride to the top. Vostok lead designer Oleg Ivanovsky helped him into the spacecraft and switched on the radio communications system. The ejection seat Gagarin was strapped into would have been of only limited use during the launch, even though ejection was the planned response to a catastrophic problem with the R-7 on the launch pad. In truth, Gagarin would never have gained sufficient altitude for the seat's parachutes to work. Because of this, a huge net designed to catch the ejector seat was positioned some 1500 metres (4900 ft) from the pad.

The officials walked back to the main command bunker. A small table with a green tablecloth had been laid out specifically for Korolev. There was a two-way radio for communicating with Gagarin in the capsule and a red telephone for giving the password to fire the rockets on the escape tower in case of an emergency during the first 40 seconds of the mission. Only three people knew the password. Gagarin's call sign was 'Kedr-Cedar', while the ground call sign was 'Zarya-Dawn'.

The flight was designed to be automatic. Ideally Gagarin would have nothing to do. But in the event of a malfunction he could take over command of the spacecraft by punching numbers into the keypad to release the controls, enabling him to use the thrusters to manually orientate it for re-entry. The numbers were in an envelope in the capsule but Oleg Ivanovsky said they were three, two and five. Gagarin replied, to Oleg's surprise, that he knew because Kamanin had already told him. Checks showed the hatch was not sealed properly. So engineers removed all 30 screws and then shut the hatch again; this time all the indicators were positive. Just as they finished, the gantry started to retract automatically towards its 45 degree angle for launch with the engineers still on it. A frantic phone call to the control room stopped the retraction for a few minutes while they descended. They finally left the vicinity about 30 minutes prior to the scheduled launch.

Ground Control: *Yuri. You're not getting bored there, are you?*

Gagarin replied: *If there was some music, I could stand it a little better.*

Ground Control: *One minute. Station Zarya, this is Zarya. Fulfil Kedr's request. Give him some music. Give him some music. Did you read that?*

Launch of the *Vostok 1* spacecraft carrying Yuri Gagarin into space.

At T–15 minutes Gagarin put on his gloves, and ten minutes later he closed his helmet. Korolev took tranquillizer pills. Of 16 launches involving this rocket, eight had failed. Of the seven Vostok spacecraft flown, two had failed to reach orbit because of booster malfunctions, while two others had failed to complete their missions. There was no American-style three, two, one countdown – just a checklist, which was soon completed.

A Historic Flight

At 9.06 and 59.7 seconds on the morning of 12 April 1961, the Vostok spacecraft lifted off with its 27-year-old passenger. 'We're off,' he cried. Korolev had the abort codes ready in case the booster did not achieve normal performance, but the launch trajectory was on target. After 19 seconds, the four strap-on boosters separated. The capsule's shroud broke away 50 seconds later. At about 5 G, Gagarin reported some difficulty in talking, saying that all the muscles in his face were drawn and strained. The G-load steadily increased until the central core of the launcher ceased to operate and was detached at T+300 seconds. Gagarin's pulse reached a maximum of 150 beats per minute.

Korolev was visibly shaking as the dramatic event proceeded. Incoming telemetry began to stream in a series of 'fives', indicating all was well. Then they changed to 'threes'. There were brief seconds of terror – a 'two' was a malfunction, but what was a 'three'? After a few agonizing moments, the numbers reverted back to 'fives'. Feoktistov remembers that: 'these interruptions, a few seconds in length, shortened the lives of the designers.'

> Gagarin: *I see the Earth. The G-load is increasing somewhat. I feel excellent, in a good mood. I see the clouds. The landing site. It's beautiful. What beauty. How do you read me?*
>
> Ground Control: *We read you well. Continue the flight.*

Orbital insertion occurred at T+676 seconds just after shutdown of the third-stage engine. The orbit was much higher than had been planned for the flight; the apogee – the furthest point from the Earth in the spacecraft's orbit – was about 43.5 miles (70 km) over the planned altitude, indicating a less than optimum performance by the rocket. Korolev had been right to worry.

Gagarin reported that he had been in good shape:

> *I ate and drank normally. I could eat and drink. I noticed no physiological difficulties. The feeling of weightlessness was somewhat unfamiliar compared with Earth conditions. You feel as if you were hanging in a horizontal position in straps. You feel as if you are suspended. Later I got used to it and had no unpleasant sensations. I made entries into the logbook, reported, worked with the telegraph key. When I had meals I also had water. I let the writing pad out of my hands and it floated together with the pencil in front of me. Then when I had to write the next report. I took the pad but the pencil wasn't where it had been. It had flown off somewhere.*

Once the orbit had been determined, the data was sent to Moscow and reporters were instructed to open their secret envelopes. It took the Soviet news agency TASS an hour to broadcast the news:

> *The world's first satellite-ship 'Vostok' with a human on board was launched into an orbit about the Earth from the Soviet Union. The pilot-cosmonaut of the spaceship satellite 'Vostok' is a citizen of the Union of Soviet Socialist Republics. Major Yuri Alexeyevich Gagarin.*

The Americans already knew. A radio surveillance station in Alaska had detected transmissions from the spacecraft 20 minutes after launch.

A View from Space

The capsule was spinning slowly, and through the porthole Gagarin could see the blackness of space and the blue-white of the Earth beneath him. He could not see the stars. The television camera trained on his face required a bright light that almost dazzled him. He said: 'I can see the clouds, everything. It's beautiful.'

Vostok's path took it over Siberia, up to the Arctic Circle, across the Kamchatka Peninsula and into the Earth's shadow over the Pacific. As Vostok's orbit took it over Cape Horn and into the South Atlantic on the final leg of its journey, it was time to prepare for re-entry. Seventy-nine minutes after lift-off Vostok automatically oriented itself, then the retro-rocket system fired for 40 seconds at 10.25 am. As soon as the braking rocket cut out, there was a sharp jolt, and Vostok began to rotate very quickly.

> Gagarin: *I had barely enough time to cover myself to protect my eyes from the Sun's rays. I put my legs to the porthole, but didn't close the blinds.'*

There had been a serious malfunction. The large instrument section of the vehicle was due to separate from the spherical descent capsule but it did not happen. 'I wondered what was going on and waited for the separation. There was no separation,' Gagarin said later. The mechanism detached the two modules as planned but the compartments remained loosely connected by a few cables. It was serious but not life threatening, and the instrument section did in fact break off later. Gagarin reported: 'I used the telegraph key to transmit the 'VN' message meaning "all goes well".'

During re-entry Gagarin saw a bright purple light at the edges of the blinds and said he felt the capsule oscillate and the coating burn away with cracking sounds. He was subjected to an intense 10 G and for about two or three seconds the instrument readings became blurred. 'My vision became somewhat greyish. I strained myself again. This worked,' Gagarin said. At an altitude of 7000 metres (23,000 ft) parachutes opened, and then the hatch was jettisoned. Gagarin was ejected a few seconds later and, looking down, recognized he was near the Volga. He separated from his seat, and his personal parachute deployed.

Touchdown

Ground control spent several anxious minutes when communications were cut off soon after the retrorockets fired. Korolev telephoned Khrushchev, who was at the holiday resort at Pitsunda: 'The parachute has opened, and he's landing. The spacecraft seems to be OK!' Khrushchev begged to know:

Is he alive? Is he sending signals? Is he alive? Is he alive?

At 10.55 am, just one hour and 48 minutes following launch, Gagarin landed softly in a field next to a deep ravine 18 miles (29 km) southwest of the town of Engels in the Saratov region. It took him six minutes to take off his spacesuit.

> Gagarin: *I had to do something to send a message that I had landed normally. I climbed a small hill and saw a woman with a girl approaching me. She was about 800 metres (2625 ft) away from me. I walked to her to ask where I could find a telephone. She told me that I could use the telephone in the field camp. I asked the woman not to let anyone touch my parachute.*

Korolev was beside himself, laughing and smiling for the first time in days. Members of the commission flew to the landing site to inspect the capsule. Korolev did not see it until later and reportedly could not take his eyes off it repeatedly touching it. Upon seeing Korolev, Gagarin reported quietly: 'All is well, Sergei Pavlovich.'

The *New York Times* ran the headline 'Soviet Orbits Man And Recovers Him'. It was a headline that echoed around the world. Gagarin returned to Moscow Airport flanked by an escort of fighter planes, while thousands of onlookers cheered him on a procession to Red Square where Khrushchev, Brezhnev and other leaders of the Soviet state basked in the unqualified triumph. Derided for years by the West for its antiquated technology, the Soviet Union had taken one of the most important steps in history. Korolev, the chief architect of this achievement, travelled several cars behind the leading motorcade and was forbidden from wearing previous state awards on his lapel for fear that Western agents might recognize him.

Shortly afterwards US President John F. Kennedy asked Vice-President Lyndon Johnson for recommendations on activities in space that would provide 'dramatic results' and beat the Soviets.

'Let's go and get the job done'

AMERICA REACHES SPACE

THE MERCURY FLIGHTS OF ALAN SHEPARD AND JOHN GLENN

1961–1962

Following the American disappointment at losing the race into space, Alan Shepard became the first US astronaut in space in a 15-minute sub-orbital flight on 5 May 1961. Barely three weeks later, President Kennedy threw down the gauntlet with a call for the landing of an American on the Moon by the end of the decade. Gus Grissom flew a similar flight to Shepard in July, followed by the Russian Gherman Titov who circled the Earth in August. In early 1962 John Glenn became the first American to orbit Earth, followed later by Scott Carpenter.

It is entirely possible that Yuri Gagarin could have been the second human to go into space, although he would have been the first to orbit the Earth. The first person in space could have been Alan Shepard. He had been waiting for his suborbital Mercury flight. Later he said:

> That little race between Gagarin and me was really, really close. Obviously, their objectives and their capabilities for orbital flight were greater than ours at that particular point. We eventually caught up and went past them, but it was the Cold War, there was a competition.

Shepard's flight was originally scheduled for 24 March, but in late January the Kennedy administration received a critical report from a government science advisory group known as the Wiesner Committee. It said there should be an immediate delay in the first manned flight, due in part to the unreliability of the *Redstone* booster. One of the committee's heads, George Kistiakowski, even declared that launching Shepard too early would provide the astronaut with: 'the most expensive funeral man has ever had.' The Wiesner Report criticized NASA's manned spaceflight programme, which placed pressure on its administrator, James Webb, and Robert Gilruth, who was in charge of manned spaceflight. Consequently, they discussed the flight and told Wernher von Braun and his rocket team that a further unmanned test

The first American to fly in space, Alan Shepard
waits, fully prepared in his pressure suit, to be
loaded into *Freedom 7* on 5 May 1961.

1961

25 April *Mercury-Atlas* rocket lifts off with an electronic mannequin

5 May Astronaut Alan Bartlett Shepard Jr becomes the first American in space, making a 15-minute suborbital flight in the *Freedom 7* Project Mercury capsule

25 May President John F. Kennedy makes a pledge to put a man on the Moon by the end of the decade

21 July Captain Virgil 'Gus' Grissom becomes the second American in space, flying in the Mercury *Liberty Bell 7*

6 August Russian cosmonaut Gherman Titov circles the Earth 17 times in a 25-hour flight in *Vostok 2*

13 September An unmanned *Mercury* capsule orbits the Earth and is recovered by NASA in a test for the first manned Mercury flight

27 October The first Saturn launch vehicle makes an unmanned flight

29 November Chimpanzee makes orbital flight in the Mercury capsule

1962

20 February Lieutenant Colonel John H. Glenn Jr becomes the first American to orbit the Earth, making three orbits in *Friendship 7*

24 May Scott Carpenter aboard *Aurora 7* becomes the second American to orbit the Earth

flight, a so-called 'booster development launch', would have to be made on the date originally set for Shepard's flight. If it were successful then he would fly on 25 April. Von Braun, who had wanted another test of the *Redstone*, agreed. Whatever the justification, it cost the United States the prize of sending the first person into space. Nineteen days later, a triumphant Soviet Union successfully put the first man into space. The news both shattered and infuriated Alan Shepard. According to the astronauts' nurse, Dee O'Hara: 'Gagarin's flight made us look like fools. Alan was bitterly disappointed, and I could understand that.'

America's First Astronaut

Alan Bartlett Shepard Jr has perhaps the most remarkable story to tell of all American astronauts. He could trace his New England ancestry back through eight generations to the *Mayflower*. Born on 18 November 1923 in Derry, New Hampshire, the son of a banker, he graduated from the US Naval Academy in 1944 and saw action in the Pacific aboard the destroyer *Cogswell*. After the war he gained his aviator's wings and went on to become a test pilot before his eventual selection as one of the Mercury 7.

Three weeks after Gagarin's flight, Shepard finally had his chance, as he recalls:

The countdown had been running very, very, well. The Redstone rocket checked out well. We had virtually no problems at all and were scheduled for, I believe it was, the second of May. And I was dressed, just about going out the door, when a tremendous rainstorm – thunderstorm – came over and obviously they decided to cancel it, which I was pleased they did. It was rescheduled for three days later, and of course, went through the same routine. The weather was good, and I remember driving down to the launching pad in a van which was capable of providing comfort for us with a pressure suit on and any last-minute adjustments in temperature devices and so on that had to be made; they were all equipped to do that.

We pulled up in front of the launch pad, of course, it was dark. The liquid oxygen was venting out from the Redstone. Searchlights all over the place. And I remember saying to myself: 'Well, I'm not going to see this Redstone again.' And you know, pilots love to go out and kick the tyres. It was sort of like reaching out and kicking the tyres on the Redstone *because I stopped and looked at it, looked back and up at this beautiful rocket, and thought: 'Well, okay buster, let's go and get the job done.' So I sort of stopped and kicked the tyres then went on in and on with the countdown.*

There was a time during the countdown when there was a problem with the inverter in the Redstone. *Gordon Cooper was the voice communicator in the blockhouse. So he called and said: 'This inverter is not working in the* Redstone. *They're going to pull the gantry back in, and we're going to change inverters. It's probably going to take about an hour, an hour-and-a-half.' And I said: 'Well, if that's the case then I would like to get out and relieve myself.' We had been working with a device to collect urine during the flight that worked pretty well in zero-gravity but it really didn't work very well when you're lying on your back with your feet up in the air like you were on the* Redstone. *And I thought my bladder was getting a little full and, if I had some time, I'd like to relieve myself. So I said: 'Gordo, would you check and see if I can get out and relieve myself quickly?' And Gordo came back — it took about three or four minutes — and said, in a German accent: 'No,' he says; 'Wernher von Braun says: "The astronaut shall stay in the nosecone".' So I said: 'Well, all right that's fine but I'm going to go to the bathroom.' And they said: 'Well, you can't do that because you've got wires all over your body and will have short circuits.' I said: 'Don't you guys have a switch that turns off those wires?' And they said: 'Yeah, we've got a switch.' So I said: 'Please turn the switch off.' Well, I relieved myself and of course with a cotton undergarment, which we had on, it soaked up immediately in the undergarment and with 100 percent oxygen flowing through that spacecraft, I was totally dry by the time we launched. But somebody did say something about me being in the world's first wetback in space.*

Finally Shepard was able to say:

Roger lift-off and the clock is started. This is Freedom 7, *the fuel is go. 1.2 G, cabin at 14 psi. Oxygen is go.*

The capsule separated from the *Redstone* five minutes later and carried on upwards to reach an altitude of 116 miles (187 km). 'OK, it's a lot smoother now,' reported Shepard. 'On the periscope what a beautiful view. Cloud cover over Florida, three to four tenths near the eastern coast … I can see Okeechobee. Identify Andrews Island, identify the reefs.' He experienced four minutes and 45 seconds of weightlessness.

Kennedy Sets a Challenge

Shepard: *We were invited back to Washington after the mission, and I got a nice little medal from the president, and which by the way he dropped. I don't know whether you remember that scene or not, but Jimmy Webb had the thing in a box and it had been loosened from its little clip, and so as the president made his speech and said: 'I now present you the medal,' and he turned around and Webb leaned forward, and the thing slid out of the box and went to the deck, and Kennedy and I both bent over for it. We almost banged heads. Kennedy made it first and he said,*

in his damn Yankee accent: 'Here, Shepard, I give you this medal that comes from the ground up.'
Jackie Kennedy is sitting there, she's mortified and said: 'Jack, pin it on him. Pin it on him!' So
he then recovered to the point where he pinned the medal on and everything was fine, and we had
a big laugh out of that.

Johnson presented his report to Kennedy five days after Shepard's flight, calling for an acceleration of US efforts to explore space 'to pursue projects aimed at enhancing national prestige'. Then came Kennedy's State of the Union address to a joint session of Congress on 25 May 1961.

I believe that this nation should commit itself to achieving the goal, before this decade is out, of
landing a man on the Moon and returning him safely to Earth. No single space project in this
period will be so difficult or expensive to accomplish.

Kennedy's speech was not widely reported in the Soviet media; few in their space programme took any notice.

Shepard: *Just three weeks after that mission, 15 minutes in space, Kennedy made his*
announcement: 'Folks, we are going to the Moon, and we're going to do it within this decade.'
After 15 minutes of space time! Now, you don't think he was excited? You don't think he was a
space cadet? Absolutely, absolutely! People say: 'Well, he made the announcement because he had
problems with the Bay of Pigs, his popularity was going down.' Not true! When Glenn finished his
mission, Glenn, Grissom and I flew with Jack back from West Palm to Washington for Glenn's
ceremony. The four of us sat in his cabin and we talked about what Gus had done, we talked
about what John had done, we talked about what I had done. All the way back. People would
come in with papers to be signed and he'd say: 'Don't worry, we'll get to those when we get back
to Washington.' The entire flight. I tell you, he was really, really a space cadet. And it's too bad he
could not have lived to see his promise.

A Second American in Space

Two months after Alan Shepard's flight, Gus Grissom performed a similar mission in the *Liberty Bell 7* Mercury capsule. It was a straightforward flight until splashdown, when the explosive bolts securing the hatch blew prematurely. There is speculation that Grissom had accidentally armed the bolts when his elbow pressed against a switch. Grissom then had to make a quick exit as *Liberty Bell 7* started to sink. As he had not sealed the hose connection on his spacesuit, it started to fill with sea water. Meanwhile, the capsule was tethered to the helicopter but it was also filling with water and getting heavier. A warning light in the helicopter indicated that it was about to flounder and so the pilot ejected the capsule, which sank 3 miles (4.8 km) to the Atlantic Ocean floor, where it remained for 38 years before it was retrieved.

It was later discovered that the helicopter's warning light was a false alarm and that the capsule could have been recovered. The capsule is now on display in the Kansas Cosmosphere and Space Center. There were no clues in the capsule as to why the hatch blew.

Titov Gets His Chance

In the Soviet Union, although there were orders for the manufacture of more Vostok spacecraft, detailed plans for future missions were rather vague. Unlike the United States, which had a specific series of missions and goals as part of its Mercury project, the Soviet effort was moving forward in a rather haphazard way. It was to be their undoing. Vague plans for the second piloted Vostok flight dated back to early 1961, focusing on a daylong mission.

Titov, Gagarin's backup for the first mission, was a natural choice for the flight. For Titov's backup, the most likely choice would have been Nelyubov, but Titov apparently had been irritated by Nelyubov's outspoken attitude so he was dropped and Nikolayev became the backup. Three months after Gagarin's flight, Khrushchev invited Korolev and a number of other prominent space figures to meet with him on a vacation in Crimea. Korolev said that a second Vostok mission was in preparation. Khrushchev added that the launch should occur no later than 10 August. Later the reason became clear – the building of the Berlin Wall began on 13 August. Khrushchev had wanted to give the socialist world a morale boost during such a tense time.

As the launch date approached there was some trepidation because of increased radiation resulting from intense solar activity, but it declined in time. On the morning of 6 August, Titov blasted off. This time the booster worked as expected, but when Titov entered orbit he was not well. He felt as he was flying upside down and in a 'strange fog', unable to read the instrument panel. On the second orbit he felt worse and thought of asking that the flight be curtailed. He tried eating a little but vomited. He carried out an experiment manually, firing the attitude control jets, but although it went well he still felt terrible, only slightly better than in previous orbits.

Now I'm going to lie down and sleep. You can think what you want, but I'm going to sleep.

Flight rules said he had to keep his helmet on when sleeping but he felt that he might choke if he vomited. So he rigged up a piece of string to jerk open the visor in case of an emergency while sleeping. He overslept by about 30 minutes, waking on his 12th orbit, at the end of which he began to improve. The re-entry was as complicated as it had been on Gagarin's mission; the instrument section again remained attached to the spherical descent capsule. Eventually Titov ejected after a record flight of one day, one hour and 11 minutes.

An American in Orbit

In November NASA fished chimpanzee Enos from the Atlantic after he had made the second orbital test of the Mercury capsule. The third suborbital *Redstone* flight was cancelled and the next one would be the more powerful *Atlas* booster taking John Glenn into orbit. On 20 February 1962, after several attempts, it finally took off. The *Atlas* booster was unusual in its use of balloon tanks for holding the fuel; these were made from very thin stainless steel with minimal or no rigid support. It was pressure in the tanks due to the fuel that provided the structural rigidity required for flight. In fact an *Atlas* rocket would collapse under its own weight if not kept pressurized. It needed nitrogen in the tank even when not fuelled.

Astronaut John Glenn climbs into his spacecraft *Friendship 7*
on 20 February 1962, prior to his launch on the first
American manned orbital mission.

Five minutes after the launch on 20 February Glenn reported:

*Capsule is good. Zero G and I feel fine. Oh that view is tremendous. The capsule is turning
around and I can see the booster during turnaround just a couple of hundred yards behind me.
It is beautiful.*

Mission Control: *Roger, Seven. You have a go at least seven orbits.*

Twenty-seven minutes after launch Glenn traversed the Sahara. He could see some fires down on the
ground, long smoke trails right on the edge of the desert. Over the Indian Ocean on the first orbit:

*The sunset was beautiful. It went down very rapidly. I still have a brilliant blue band clear across
the horizon almost covering my whole window.*

After 50 minutes concerning orbital night:

> *The only unusual thing I have noticed is a rather high, what would appear to be a haze layer, up some seven or eight degrees above the horizon on the night side. The stars I can see through it as they go down towards the real horizon, but it is a very visible single band or layer pretty well up to the normal horizon. This is Friendship 7. I have the Pleiades in sight out here, very clear.*

Then Glenn saw something unusual outside *Friendship 7*:

> *I am in a big mass of some very small particles, they're brilliantly lit up like they're luminescent. I never saw anything like it. They round a little: they're coming by the capsule, and they look like little stars. A whole shower of them coming by. They swirl around the capsule and go in front of the window and they're all brilliantly lighted.*

A Potential Disaster

On the third orbit a potentially serious situation occurred. One of the sensors on board *Friendship 7* indicated that the spacecraft's heat shield was loose. If the heat shield was to break away during re-entry, Glenn would perish.

> '*This is Friendship 7. I have the Pleiades in sight out here, very clear.*'
>
> JOHN GLENN

Gene Kranz in Mission Control at Cape Canaveral at the time remembers the incident well as a key moment in the evolution of mission operations:

> *What the crew was seeing, we were seeing on board the spacecraft in Mercury. It was only as we moved into Gemini that we recognized the need to move deeper into the spacecraft system. Part of this came about as a result of the John Glenn mission. Because in John Glenn, we were stuck with a very difficult decision. Did his heat shield deploy or did it not? We had a single telemetry measurement that indicated that the heat shield had come loose from the spacecraft. Now, if we believed that measurement and the heat shield had come loose, we had one set of decisions that involved sticking our neck out by retaining the retrorocket package attached during the entry phase. We didn't know whether it would damage the heat shield. We didn't know whether we had sufficient attitude control authority. So if the heat shield had come loose and we believed that measurement, we'd go that direction. But if the heat shield had not come loose, that measurement was wrong and we wouldn't do anything different. So it was a very difficult decision.*
>
> *I remember this one very clearly, because the engineers would come and say: 'Nah, the heat shield can't have come loose!' And Chris Kraft would look at them, and he'd say: 'Well, how about this measurement we're seeing? What's the worst thing that would happen if it had come*

loose?' And they'd always end up in a position that says: 'Well, maybe John Glenn isn't going to make it home.' 'Well then, what are we going to do about it?' So, because of Kraft, this entire business of ground control, I think, really came into being on the Mercury-Atlas 5.

The Mercury capsule was designed by Max Faget, a brilliant engineer who had been working with NASA since its inception. He remembers:

Well, this is one thing that I'm kind of proud of, in a way. We considered all possible kinds of failures, and one of the failures that we considered was that the retropack would not jettison. The retropack sat on the heat shield, and it was held by three straps. I thought, well, maybe the damned thing wouldn't get jettisoned for some reason. So we ran some wind tunnel tests with the retropacks on there. We found out it was stable with the retrorockets on there. So we had these sensors, and the sensors were designed so that if any one sensor said that they had not been opened, it would let you know that. Any sensor that opened would let you know it had been done.

Apparently we had a bad sensor on that flight, and we thought that the heat shield had been released. Well, the arrangement was such that if the heat shield were released, the straps would have still held it on. So everybody was concerned that the heat shield had been released, because that's what the instrumentation said. Well, they called me up – I was back in Houston – and asked me what about it, and I said: 'Well, you can enter that way because we've got wind tunnel data that said the thing will be stable,' and they did.

Now they had to inform John Glenn on board *Friendship 7*. Initially they did not tell him the whole story, however.

Mission Control: *This is Texas Cap Com,* Friendship 7. *We are recommending that you leave the retropackage on through the entire re-entry.*

Glenn: *This is* Friendship 7. *What is the reason for this? Do you have any reason? Over.*

Mission Control: *None at this time. This is the judgement of Cape Flight.*

An hour later, a few minutes before re-entry, they told him:

We are not sure whether or not your landing bag has deployed. We feel it is possible to re-enter with the retropackage on. We see no difficulty at this time with that type of re-entry. Over.

Re-entry was a dramatic affair.

Glenn: *My condition is good, but that was a real fireball. Boy. I had great chunks of that retropack breaking off all the way through.*

A Hero Returns

Glenn became a national hero and had one of the biggest New York tickertape welcomes ever. But it seemed that the flight of *Friendship 7* would be his last:

> Well, after my flight I wanted to get back in rotation and go up again. Bob Gilruth, who was running the programme at that time, said that he wanted me to go into some areas of management of training and so on, and I said I didn't want to do that. I wanted to get back in line again for another flight. But he said headquarters wanted it that way, at least for a little while. And I didn't know what the reason for this was, and I kept going back. Every month or two I'd go back and talk to him again about when do I get back in rotation again, and he'd tell me: 'Well, not now. Headquarters doesn't want you to do this yet.'
>
> I don't know whether he was afraid of the political fallout or what would happen if I got bagged on another flight. I don't know what it was, but apparently he didn't want me used again right away. So that's the reason I never got another flight. Bob Gilruth kept saying, well, that he wanted me here in training and management, plus the upcoming flights should go to people who would be useful for the early lunar landings, and that by the time those were expected to occur, I'd be over 50, and that might be a little too old.

But by a curious twist of fate John Glenn was to get into space again, 36 years later!

Aurora 7

Scott Carpenter, John Glenn's backup for the first US orbital flight, went into space in May. When the prime pilot for the mission after Glenn's, Deke Slayton, was grounded because of a heart flutter, it was Carpenter and not Slayton's backup, Wally Schirra, who got the flight. It was bad luck for Slayton who, had the original flight schedule been adhered to, would have become the first American to orbit the Earth. The flight of *Aurora 7* has been unfairly criticized as a poor flight. It was planned to be a scientific mission as well as conducting the most thorough workout of the Mercury capsule to date. However, various malfunctions resulted in the need for frequent attitude corrections and a heavy use of fuel. Carpenter also made a serious error by not turning off the automatic orientation system when he switched to manual control prior to re-entry, which resulted in a critical waste of fuel. Further malfunctions and the low level of fuel meant that when Carpenter did align *Aurora 7* correctly for retrofire, it was five seconds too late. Re-entry itself was dramatic, with the capsule oscillating wildly. It overshot the landing zone by 250 miles (402 km). Carpenter left NASA shortly afterwards.

VALENTINA TERESHKOVA – VOSTOK 6
1962–1963

During late 1962 and early 1963, in an attempt to regain the headlines, the Russians planned a number of high-profile missions: the first multiple mission and the longest flight. The most eye-catching was putting the first woman into space – this was achieved on 16 June 1963 with the flight of Valentina Tereshkova. Meanwhile the Americans impressively flew two more Mercury missions, completing the programme at six manned missions.

The future plans of the Soviet Union depended upon Korolev's unwritten rule that each mission be a significant advance over the previous one, although there was no overarching plan. One month after Titov's troubled flight, Korolev proposed a dramatic mission – three Vostok spacecraft, each with a single cosmonaut, to be launched on three successive days. The first pilot would conduct a three-day mission, while the two others would be in space for two or three days. On one of those days, all three spacecraft would be in space. But others were not convinced about the viability of the project, and Korolev was forced to reduce the plan to two Vostok craft, launched by January 1962 at the earliest.

The First Women Cosmonauts

Meanwhile, the Central Committee had approved the hiring of 60 new cosmonaut trainees, including five women: Tatyana D. Kuznetsova, aged 20; Valentina L. Ponomareva, aged 28; Irina B. Solovyeva, aged 24; Valentina V. Tereshkova, aged 24; and Zhanna D. Yerkina, aged 22. Solovyeva had 900 parachute jumps to her credit, followed by Tereshkova with 78, and Ponomareva with ten. Although Ponomareva was clearly the most accomplished pilot, Gagarin opposed her inclusion because she was a mother. Another candidate, Tereshkova, did not have any academic honours but had been an active member of the local Young Communist League.

The publicity surrounding Glenn's launch had not gone unnoticed in the USSR. Military-industrial Commission Chairman Ustinov called Korolev on 7 February, just days before Glenn's flight, and ordered

Valentina Tereshkova, the first woman in space was launched on 16 June 1963, returning on 19 June.

the dual Vostok launch in mid-March. In his diary Kamanin commented on the foolishness of making decisions in such a way:

> *This is the style of our leadership. They've been doing nothing for almost half a year and now they ask us to prepare an extremely complex mission in just ten days' time. The programme of which has not even been agreed upon.*

Fortunately a rocket failure at Baikonur forced a much-needed delay to the dual Vostok mission.

A Surprise Second Launch

Cosmonauts Nikolayev and Popovich were the obvious candidates for the two missions. One of the few bachelors in the team, the 32-year-old Nikolayev began his career as a lumberjack before later joining the Soviet air force, receiving his pilot's wings in 1954. Popovich, also 32, had enjoyed a distinguished career in the Soviet air force before receiving the Order of the Red Star for an assignment in the Arctic. His wife Marina was one of the most accomplished women test pilots in the USSR.

On 11 August Nikolayev took off. Korolev was so nervous throughout the ascent phase that he held tightly to the red telephone with which he would give the order to abort the mission in case of a booster failure. Khrushchev spoke to Nikolayev four hours into the mission, and the world saw Nikolayev smile on television. As *Vostok 3* passed over Baikonur at 11.02 a.m. a day later, *Vostok 4* climbed after it. It was the first time that more than one piloted spacecraft, or indeed more than one human, had been in orbit. Western media was surprised by the second launch, speculating that there would be a docking. There was talk that the mission was a rehearsal for a Moon flight, but watchful commentators noticed that this was not a true rendezvous, just two spacecraft launched into similar orbits, neither of which could be altered. Both Vostoks fired their retrorockets within six minutes of each other on 15 August. Nikolayev landed after a three-day, 22-hour and 22-minute flight, during which he had circled Earth 64 times. Popovich landed 125 miles (200 km) away after a two-day, 22-hour and 57-minute flight, and 48 orbits.

Korolev breathed a sigh of relief. His political masters were satisfied, but his health was worsening. He had been in poor condition for many years; the effects of the privations of the labour camps had never left him. His busy work schedule aggravated matters – working 18 hours a day for several weeks on end was common. He found it hard to delegate, often involving himself in trivial matters he should have left to others. Soon after the return of the twin Vostoks he suffered intestinal bleeding. After a stay in the hospital, he was ordered to take a holiday at the seaside resort of Sochi, but he took his work with him and was constantly on the telephone.

'Fireflies' on the *Sigma 7* Mercury Flight

The Mercury programme was gaining momentum with the launch of Walter Schirra in the *Sigma 7* capsule in October 1962. He intended to fly a technically perfect mission:

Schirra: *Not to criticize John and Scott, but the mission was designed to have a chimpanzee in there. They replaced the chimp. But that meant they had to have a lot of automatic manoeuvres. Automatic manoeuvres took a tremendous amount of attitude control fuel. I said: 'I don't want to do that. I just want to save that.' And as a result, I ended up, I think, about retrofire, about 80 percent of my attitude fuel was still remaining.*

As water came out of the spacecraft it froze instantaneously into one snowflake, but a very tiny, tiny snowflake. These stuck on the outside of the spacecraft. They drifted around. This was what John called fireflies, is what Scott got involved with banging the spacecraft and watching them come off. And as a result, both of them lost sight of the fact they had to have fuel enough to fly the mission. John got a little wrapped up; I did, too, because I was his Capcom in California, on the retro-rocket package that had to be kept on because of a false signal that said his heat shield had detached, when in fact it turned out it had not. But at any rate, that became kind of a traumatic part of John's mission. But in both cases, they almost ran out of attitude control fuel; and that kind of shook me up, because there's no reason to do that. In fact, I alienated some of the flight controllers because, after drifting for a while, I put it back into automatic control. I'm in chimp mode now; it didn't go over too well.

1962

11 August *Vostok 3* launched

12 August *Vostok 4* launched to join *Vostok 3* in space

3 October *Sigma 7*, part of the Mercury programme, launched with Walter Schirra aboard

1963

15 May Launch of *Faith 7* with Gordon Cooper aboard marks the end of the US Mercury programme

14–19 June Valery Bykovsky completes the longest manned space flight to date (199 hours, 81 orbits) during the *Vostok 5* mission

16 June *Vostok 6* launched with first woman in space, Valentina Tereshkova, aboard

End of the Mercury Missions

The final Mercury flight occurred in May 1963 when the relaxed Gordon Cooper flew in his capsule *Faith 7*. He had to endure what was becoming a common problem on Mercury flights – spacesuit overheating. The flight was nevertheless going well and he remarked how much detail he could see down on the ground. He deployed a flashing beacon from the nose of his capsule to test how far he could see it – an important procedure for future rendezvous missions. Then there was trouble, as he later described:

On the 19th orbit a warning light came on. The .05 G green warning light came on, which is the light that tells you you're starting to re-enter. I was sure that I wasn't re-entering, because there had been nothing to slow down my speed at all. And, of course, as usually happened on these missions, we had long spaces when we were out of radio contact; and I was out of radio contact when this happened. So when I got in radio contact first time, the Cape was kind of concerned when they heard about this light on. Then we proceeded on the next orbit or so to try to analyse, go through various procedures to try to find what it was. And we realized I was, slowly but surely, having an electrical fire from my relays; and they did short out the inverters. So, eventually I lost my total electrical system.

It meant that I had the manual push/pull rods to activate the jets for attitude control. I had eyeballs out the window for my attitude – my pitch, roll and yaw attitude. I had a wristwatch for timing. And I had to activate each and every one of the relays, and I'd have to manually fire the retros while manually flying the spacecraft. So, everything had to be done manually. I'd have to control the spacecraft all the way through re-entry. I'd have to put my drogue out manually. And I'd have to deploy my parachute manually. I'd have to deploy the landing bag manually.

In the end it was a perfect splashdown just 4 miles (6.4 km) from the USS *Kearsarge* in the Pacific.

Mercury ended with a total of two days, five hours and 55 minutes of cumulative space time from six missions. It might not have sounded much but it was a sound start, verifying the technology necessary to maintain a human in Earth orbit for a short period of time. Well before that last flight, plans for a second-generation spacecraft were already on the drawing boards. As early as December 1961, NASA Associate Administrator Robert Seamans approved a 'Mercury Mark 2' vehicle proposed by the former Space Task Group, which had been renamed the Manned Spacecraft Center. This new spacecraft would be capable of conducting extensive rendezvous and docking operations in Earth orbit, allowing astronauts to acquire experience in techniques needed for the Apollo lunar landing programme. By January 1962, the project had been renamed Gemini. It was clearly a major leap in capabilities

Wernher von Braun (centre) explains to President John F. Kennedy the *Saturn* launch system in 1963.

S-IV

UNITE

over either Mercury or Vostok. The spacecraft would be able to change orbits, it would carry two astronauts and it would allow flights lasting as long as two weeks.

Alan Shepard Grounded

Wernher von Braun had transferred to NASA from the army in 1960, having extracted the proviso that NASA develop the powerful rocket needed to launch astronauts to the Moon. He had already produced the *Jupiter-C*, so they named the new rocket after the next planet out from the Sun – Saturn.

Alan Shepard was chosen to fly in the Gemini programme but a medical problem grounded him and that seemed, at the time, to be the end of his career as an astronaut:

> *I was chosen to make the first Gemini mission. Tom Stafford, who is a very bright young guy, was assigned as copilot, and we were already into the mission, already training for the mission. We had been in the simulators, as a matter of fact, several different times. I'm not sure whether we'd looked at the hardware in St Louis or not prior to the problem, which I had. The problem I had was a disease called Menière's; it is due to elevated fluid pressure in the inner ear. They tell me it happens in people who are Type A, hyper, driven, whatever. Unfortunately, what happens is it causes a lack of balance, dizziness, and in some cases nausea as a result of all this disorientation going on up there in the ear. It fortunately is unilateral, so it was only happening with me on the left side. But it was so obvious that NASA grounded me right away, and they assigned another crew for the first Gemini flight.*

The Russians Sack Three Cosmonauts

On 27 March 1963 three trainee cosmonauts, Nelyubov, Anikeyev and Filatev, were returning to the training centre after an evening out in Moscow. They had been drinking and became involved in an altercation with a military patrol at a railway station. Nelyubov threatened to go over the head of the offended officers if they filed a formal report against the cosmonauts. Later, officials at the Cosmonaut Training Centre requested the duty officer not to file a report against the three men. He agreed, on condition that they apologize for their behaviour. Although Anikeyev and Filatev agreed to do so, Nelyubov refused, and so the offended duty officer filed a report against the three men, and within a week they were all dismissed from the cosmonaut team. Nelyubov was one of the brightest and most qualified cosmonauts; he had served as Gagarin's second backup during the first Vostok mission, and he certainly would have gone into space in the near future.

The Race to Regain Russian Superiority

How was the USSR to respond to the impressive Mercury flights and demonstrate its superiority? It was decided that the next flight of the Vostok was to include a woman. Reports on the candidates stated that Ponomareva had the most thorough preparation and was more talented than the others:

> *She exceeds all the rest in flight, but she needs a lot of reform as she is arrogant, self-centred,*

exaggerates her abilities, and does not stay away from drinking and smoking. Solovyeva is the most objective of all, more physically and morally sturdy, but she is a little closed off and is insufficiently active in social work. Tereshkova is active in society, is especially well in appearance, makes use of her great authority among everyone who she knows. Yerkina has prepared less than well in technical and physical qualities, but she is persistently improving and undoubtedly she will be a rather good cosmonaut.

The three female Russian cosmonauts selected for
Vostok 6: (from left to right) Valentina Ponomareva and Irina
Solovyeva (both backups), and Valentina Tereshkova
– who was ultimately to be the first woman to fly in space
on 16 June 1963.

The report then reached a conclusion:

We must first send Tereshkova into space flight, and her double will be Solovyeva.

It was said among the trainees that Tereshkova was 'Gagarin in a skirt!'

The flight was set for August 1963, but then Korolev discovered a problem. His engineers realized that the operational lifetime of both the proposed spacecraft was due to expire in May–June 1963, well before the August flight, and there was no possibility of extending their 'shelf life'. They had to either launch them or scrap them, so they changed the timetable. The first spaceship, launched in May or June, would carry a man into orbit for a full eight days, while the second would carry the first woman into space for two to three days. The choice for the first mission was Valeri Bykovsky; the decision on the woman was more difficult. According to later reports, what swung the decision in Tereshkova's favour was the fact that some were lobbying too enthusiastically for Ponomareva. Without such support she would have been the first woman in space. In her book, *The Female Face of the Cosmos*, Ponomareva later wrote:

Korolev started with me: he asked why I was sad and whether I would resent it if I do not fly. I rose and said with emphasis: 'Yes, Sergei Pavlovich, I would resent it very much!' Pointing his index finger at me, Korolev said: 'You are right, you fine girl. I would have resented it too.' He spoke with emphasis, very emotionally. Then he has kept silent for a while, gave every one of us a long attentive look, and said: 'It's all right, you'll all fly into space.'

The session of the State Commission on 21 May 1963 was short, and there was no miracle. It was announced that Valentina Tereshkova was appointed the commander of the space ship, and Irina Solovyeva and Valentina Ponomareva were the backups. As I remember the physician Karpov's explanation, two backups, instead of one as for men, were appointed 'with the consideration of the complexity of the female organism'.

The Tricky Launch of *Vostok 5*

But for *Vostok 5* trouble began soon after Bykovsky arrived at the pad. Neither of the shortwave transmitters on *Vostok 5* were working, later there was a problem with the ejection hatch and then there was a control failure in the third stage. Engineers moved in. They had just six hours to repair the faults, otherwise the launch would have to be cancelled and the Vostoks would exceed their design lives. Finally, the task was accomplished but the problems were not over. In the final minute of Bykovsky's countdown a light indicated that the rocket had not severed its umbilical electrical connection to the pad. Korolev looked on the verge of panic, but all around him said they should launch. In the end the rising rocket tore the cable from its socket and left it flailing on the pad.

The First Woman in Space

Tereshkova and her backup Solovyeva were prepared for the second mission. All seemed to go well and Tereshkova lifted off two days later, becoming the first woman in space. The Vostoks flew closest to each other immediately after launch, when they passed at a distance of about 3 miles (5 km). Bykovsky later reported that he had not spotted *Vostok 6*, while Tereshkova thought she might have glimpsed *Vostok 5*. They established radio contact shortly afterwards, and within three hours of the launch Moscow television was showing live shots of Tereshkova in her capsule. But she was not feeling well; subsequent transmissions showed her tired and looking weak. She initially failed to perform one of the major goals of her mission, the manual orientation of her spacecraft. Kamanin ordered Gagarin, Titov and Nikolayev to radio new instructions. Eventually she accomplished the task, showing that if the automatic system failed she would be able to put the craft into the correct orientation for re-entry. Bykovsky reported there had been a knock, and that this had caused consternation on the ground. When questioned further, Bykovsky clarified what he had said: 'There had been the first space stool.' the Russian word for 'stool' (*stul*) had been mistaken for the word for 'knock' (*stuk*). It was a historic moment of sorts – the first time a human had made a bowel movement in space.

Tereshkova landed without incident, although she bruised her face. During Bykovsky's re-entry, like that of Gagarin and Titov, his instrument compartment failed to separate from the descent capsule as planned – it was getting to be a persistent problem. Fortunately he landed without too much worry.

It was heralded as a triumph for the Soviet Union; a woman had flown in space for longer than all the six American Mercury flights combined. Bykovsky claimed the world duration record for a single-crew spaceship; it still stands today, over 45 years after his mission.

'Friends! Before us is the Moon'

TEMPTING FATE

THE DANGEROUS VOSKHOD MISSIONS
1964–1965

After the Vostok missions, the Soviet space programme became bogged down by incompetent planning, delayed schedules and a desire to produce space 'spectaculars' – instead of developing the next phase of exploration using their Soyuz spacecaft. This strategy put lives at risk and also damaged the Russians' chances of putting a human on the Moon before the Americans.

As early as March 1963 NASA had established guidelines for performing spacewalks during the forthcoming Gemini programme. By January 1964 officials at Houston had completed the final details of the plan. The flight of *Gemini IV*, then scheduled for February 1965, would include an astronaut opening the hatch and standing up for a short period. Perhaps the US would even carry out the first spacewalk, or EVA (Extra Vehicular Activity) on this mission.

The Order is 'Three Cosmonauts!'
The successor to the increasingly outmoded Russian Vostok capsule was to be the Soyuz capsule. It was designed with a voyage to the Moon in mind, but by early 1964 it was clear to Korolev that it would not be ready by late 1964, or even early 1965. The Communist Party and the USSR Council of Ministers had already committed itself to the Soyuz in a joint decree on 3 December 1963, with its ultimate goal of a manned flight around the Moon.

With the two-man Gemini flights just months away, Soviet space officials were faced with a difficult situation. None of the four projected Vostok missions planned for 1964 would compare favourably with a Gemini flight. They were all designed to carry only a single cosmonaut; none of them included a spacewalk and none of them would have the capability of changing orbits. The Soviets were paying for their lack of organization, departmental rivalry and wasteful dilution of effort and use of space for political statements. In this climate an unlikely idea emerged, as audacious as it was dangerous.

Where the idea originally came from is lost in time. Some say that Khrushchev called Korolev and ordered him to convert the one-man Vostok spacecraft into a vehicle capable of carrying not two, but three, cosmonauts. According to Kamanin, Korolev was not pleased to receive such an order, as he recalled:

The Moon, photographed from just outside Earth's atmosphere.

It was the first time that I had seen Korolev in complete bewilderment. He was very distressed at the refusal to continue construction of the Vostoks and could not see how to re-equip the ship for three in such a short time. He said it was impossible to turn a single-seater ship into a three-seater in a few months.

A Fateful Decision

However, the account given by Khrushchev's son differs. He maintains that it was Korolev's idea. It is certainly true that Korolev was considering a three-seater Vostok as early as February 1963, and he certainly had a pathological desire to beat the Americans at all costs. But wherever it came from, the decision to upstage Gemini proved to be one of the most disastrous decisions in the history of the Soviet space effort, ignoring any natural progression of space vehicles and the resulting gain in knowledge and inserting instead a showy diversion. It was the very antithesis of what the Americans were doing – an incremental acquisition of abilities and technology. For the Soviets, the space race had degenerated into little more than a circus act of one-upmanship. Ultimately it cost them the Moon.

While the Soviet Union was engaged in falsely presenting the image of a nation at the cutting edge of space exploration, the Soyuz programme – the real future of the Soviet space effort – was put on hold. They called the new project Voskhod (Sunrise) hoping no one would realize that it was a strained and stretched Vostok packed with three worried cosmonauts.

Vostok Becomes Voskhod

Konstantin Feoktistov, the resourceful engineer who played a critical role in the design of the Vostok, was on the Voskhod design team. Adding to the view that it might have been Korolev's idea after all, he later recalled how Korolev neutralized internal opposition:

We argued that it would be unsafe, that it would be better to be patient and wait for the Soyuz space ship to be built, but in the end, of course, Korolev got his way. In February 1964 he outwitted us. He said that if we could build a ship based on the Vostok design, which could carry three people, then one of those places would be offered to a staff engineer. Well, that was a very seductive offer and a few days later we produced some sketches. Our first ideas were accepted.

> '*He said it was impossible to turn a single-seater ship into a three-seater in a few months.*'
>
> NIKOLAI KAMANIN

Feoktistov proposed getting rid of the ejection seat and spacesuits from the Vostok, thus allowing three men to cram into the spherical capsule in regular clothing. Many objected to this move but it was really a foregone conclusion; it would have been impossible to fit them in any other way. By the time the draft plan was completed it was also clear that there would not be a tower-equipped launch escape system ready for the Voskhod launch, but Korolev and his engineers took the risky step of moving on with the launch despite this blatant disregard for safety. It was stated that it would be 'difficult' to rescue the cosmonauts up to the first 25 to 44 seconds of a launch. That was wrong. If a failure occurred during that period, the crew would be doomed.

Korolev's health continued to decline. In February 1964 he suffered a heart attack and spent several days in hospital. Doctors had prescribed a long holiday, which was delayed by urgent work. He was allowed to fly to Czechoslovakia for a brief holiday, the only time between 1947 and his death that he left the Soviet Union. Upon returning to Moscow he immersed himself in the Voskhod preparations. A drop test with an engineering version of the capsule was carried out in September. It was a disaster; the parachute hatch failed to open and the capsule was smashed to pieces.

Who Flies the Risky Mission?

Then there was the question of which engineer was to fly on the mission. Feoktistov knew more about the design of the Vostok and Voskhod spacecraft than anyone, but he was not considered fit enough to be a cosmonaut. When Kamanin heard that Feoktistov was an option he was reported to have blurted out:

> *How can you put a man into a space ship if he is suffering from ulcers, nearsightedness, deformation of the spine, gastritis, and even has missing fingers on his left hand?*

The air force objected as well, but Korolev backed his engineer and they eventually capitulated. Korolev yelled in frustration:

> *The air force is perpetually jamming up the works! Looks like I'm going to have to train my own cosmonauts.*

The launch was set for the morning of 12 October. Korolev was more nervous and irritable than anyone had ever seen him. The three cosmonauts arrived at the launch pad at 10.15 a.m. local time dressed in lightweight grey woollen trousers, shirts and light blue jackets. Korolev and Gagarin saw the three men up

1964

February Soviet rocket scientist Sergei Korolev suffers a heart attack

28 July US probe *Ranger 7* is launched towards the Moon and sends back 4308 television pictures

12–13 October The first three-member Soviet crew orbits Earth on board a Voskhod spacecraft

1965

18 March Cosmonaut Alexei Leonov conducts the world's first spacewalk during *Voskhod 2* mission

6 April The US launches *Intelsat I*, also known as the 'Early Bird' communications satellite

15 July US space probe *Mariner 4* completes flyby of Mars

16 July First Soviet *Proton* rocket blasts off from Site 81 in Baikonur

to the elevator before they removed their jackets and boots, donned slippers and entered the spacecraft: Boris Yegorov first, then Feoktistov, followed by Commander Vladimir Komarov. The tension was higher than perhaps during any other mission since Gagarin's. Without a viable launch escape system during the first minute of the mission, there was absolutely no way of saving the crew in the event of booster failure. Korolev was so nervous he was shaking uncontrollably.

To his immense relief *Voskhod 1* got into orbit without a flaw. Once again, the reaction from the West was unprecedented, prompting another speculation that the ultimate Soviet plan was to go to the Moon. Within two to three hours of the launch Feoktistov and Yegorov began to experience disorientation but despite this the short mission proceeded without much incident. When they landed there was relief all round. Because they had no room for three ejector seats a solid-fuel braking rocket was added to cushion the impact of the capsule with the ground. The flight had lasted one day, 17 minutes and three seconds and achieved nothing except propaganda. They had been lucky to get away with such a gamble.

Khrushchev Loses Power

It was later that day that Korolev and Kamanin heard of the changes back in Moscow. News had come in that there would be a special meeting of the Central Committee the same evening. Within hours Khrushchev was no longer in power and had been replaced in his two posts by Alexei Kosygin and Leonid Brezhnev. Kamanin was instructed to alter the cosmonauts' speeches. Instead of saluting Khrushchev, they would salute Brezhnev and Kosygin.

A Dangerous Excursion

It was clear to Korolev that the second Voskhod mission should include a spacewalk. NASA's announced plans to carry out an EVA during the Gemini programme once again compelled him to try and beat the Americans. But how could a Voskhod capsule be modified so that a

The casually attired *Voskhod 1* cosmonauts (from left) Vladimir Komarov, Boris Yegorov and Konstantin Feoktistov on their way to the launch pad on 12 October 1964.

spacewalk could take place? Soviet engineers could not consider the Gemini method of depressurizing the entire spaceship during an EVA, because their life-support systems were not good enough and the instruments in the Voskhod capsule were not designed to operate in a vacuum. Instead, they drew up a plan for an airlock, made of rubber, to be unfolded on the outside of the spacecraft. Both cosmonauts would wear pressure suits throughout the flight. During the space walk one cosmonaut would crawl into the airlock, shut the hatch behind him, evacuate it, open an outer hatch and then step out into space. A 5-metre (16-ft) cord would connect the cosmonaut to the space ship during the EVA. The maximum time in space was limited to between ten and 15 minutes.

By this point, the best candidates for the primary crew of the *Voskhod 2* spacewalk mission were Belyayev and Leonov. The 39-year-old Pavel Belyayev had been the oldest candidate from the 'Gagarin group' of 1960. He had graduated from the Yeisk Higher Air Force School in 1945 and flew combat missions against the Japanese during the final days of the Second World War. Later, in 1959, he graduated from the famous Red Banner Air Force Academy, and thus he was only one of two cosmonauts in the 1960 class who had received a higher education. Belyayev might have flown into space earlier had it not been for a severe ankle injury sustained in August 1961 during a parachute jump, which left him out of the running for a whole year. Thirty-year-old Aleksei Leonov was born in Siberia, and graduated from the Chuguyev Higher Air Force School in the Ukraine in 1957 before serving as a jet pilot in East Germany.

Chief Designer Severin recalled:

> *The Americans planned to do their EVA in three months and had announced it beforehand. So we felt very rushed. We were hurrying and were nervous.*

The first *Voskhod 2* test spacecraft was launched into orbit successfully on 22 February 1965. Designated under the catch-all Kosmos classification, the media did not realize its true function. The fully equipped spacecraft was to simulate all the necessary airlock operations. Meanwhile, the ground tests for these aspects of the mission were beset by failures. Severin recalls:

> *The situation was really grave. Almost the entire testing programme had been disrupted. Only part of it was completed in the unmanned flight. There was even talk of postponing the flight until better results were obtained on the ground. The competition with Gemini reached such a state that Soviet security personnel arrived at Baikonur. It's possible that the KGB thought that all of our accidents were the result of sabotage. They imposed strict monitoring, which made us very nervous.*

Near Disaster

Voskhod 2 lifted off successfully and the two cosmonauts began preparations for the EVA as soon as they reached orbit. First, Belyayev expanded the Volga airlock to its full length. Then, aided by Belyayev, Leonov strapped on his life-support pack. Once the pressure between the airlock and the ship was equalized, Belyayev opened the inner hatch, allowing Leonov to crawl head-first into the airlock and hook himself up to the tether. Then Belyayev shut the inner hatch and depressurized the airlock. Leonov emerged, becoming

'I can see the Caucasus.'

ALEKSEI LEONOV

the first human to walk in space. At first, he just poked his head out, but then gradually extended his entire body. The Sun almost blinded him. His first words were: 'I can see the Caucasus.' But after 12 minutes in open space Leonov found himself in a perilous situation:

> *Near the end of my walk I realised that my feet had pulled out of my shoes and my hands had pulled away from my gloves. My entire suit stretched so much that my hands and feet appeared to shrink. I was unable to control them. I couldn't get back in straight away. My space suit had ballooned out and the pressure was quite considerable. I was tired and couldn't go in feet first as I had been taught to do.*

Leonov decreased the pressure in his suit hoping that it would make it more flexible:

> *Then I felt freer and I could move about more easily. Then I pushed myself into the airlock head first, with my arms holding the rails. I had to turn myself upside down in the airlock in order to enter the ship feet first and this was very difficult.*

His pulse raced to 143 beats per minute, his breathing was twice normal levels and his body temperature rose to 38 °C (100.4 °F). Drenched in sweat and exhausted, he closed the outer hatch behind him. Leonov had experienced depressurisation for 23 minutes and 41 seconds. The cosmonauts cast off the airlock and then settled down to a one-day mission. But there was to be yet another problem.

A Perilous Return

The hatch on the space ship had not been shut properly and was leaking air, which was being compensated for by the life-support system. The result was that the capsule was becoming rich in oxygen, which increased the possibility of a fire; a tiny spark could set off an explosion. They tried to lower the oxygen content during the remainder of their mission, bringing it down to manageable levels before re-entry. It would not be the last time that an oxygen-rich atmosphere inside a capsule would pose a risk. But for *Voskhod 2* the problems kept coming. By the 30th orbit, pressure in the cabin tanks had dropped from 75 to 25 atmospheres, bringing with it the possibility of the complete depressurization of the spacecraft. Fortunately it stabilized. When the moment for re-entry burn came around Belyayev calmly reported: 'Negative automatic retrofire.' Korolev immediately told Belyayev to use the manual system – although he was probably ahead of the craft's chief designer in thinking about how to orientate for re-entry. Once the numbered code to unlock the attitude controls was found it was handed to Gagarin who transmitted the information to Belyayev.

The exercise of orienting the spacecraft became an ordeal. They had to use an optical sighting device but both men were clad in bulky spacesuits. In the cramped space Belyayev, optical device in hand, had to lie horizontally across both seats of the capsule, while Leonov remained out of the way under his seat. At the same time, Leonov held Belyayev in place in front of the porthole so he could use both his hands to orient the ship with respect to the Earth's terminator – the boundary between day and night – using the hand controls. After this was done they quickly returned to their seats to re-establish the ship's centre of gravity before firing the retrorocket. The 46 seconds it took to get back in position before Belyayev hit the fire button resulted in a serious overshoot of their original landing point.

As with several previous Vostok missions, the instrument compartment failed to separate from the descent capsule and the two modules remained connected loosely to each other with steel straps. It resulted in a steeper than usual descent and more G forces. It burst blood vessels in both men's eyes as the load reached 10 G.

Incredibly, ignoring the obvious risk they had taken, Korolev raised a toast to the future:

> *Friends! Before us is the Moon. Let us all work together with the great goal of conquering the Moon.*

But the Soviet Union was not to launch a single manned spaceflight for two years. Voskhod was the last in a series of spectaculars, and subsequent Vostok-based missions were cancelled. The Soviet space effort had flirted with disaster. Next time they would not get away with it.

'They want you to get back in'

FLYING IN PAIRS

THE FIRST GEMINI MISSIONS
1964–1965

America's second manned space project began in April 1964 with the first flight of the Gemini series. Gemini was intended to bridge the gap between the Mercury and the Apollo programmes. Its goal was to fly two astronauts in space as well as to test equipment and techniques – such as rendezvous and docking – which were essential for the Moon missions.

In August 1965 the United States had finally taken the absolute endurance record in space with the *Gemini 5* mission, which lasted nearly a week. There were plans to fly *Gemini 7* in December for two whole weeks. In the USSR, Korolev extended the planned *Voskhod 3*'s duration from ten to 15 days and then to 20 days. Then at a meeting of the Military-Industrial Commission on 16 December 1965 the Soviet government added one more condition to the Voskhod programme: that Korolev launch two Voskhods in time for the 23rd Congress of the Communist Party in March 1966 as a salute to the party. It was an unrealistic deadline. Things were falling apart.

Gemini Begins

Although the Gemini missions were designed to carry astronauts, the first two flights involving Gemini spacecraft – *Gemini 1* (8–12 April 1964) and *Gemini 2* (19 January 1965) – were unmanned. They were used to test out the rockets and other systems prior to launching with a two-man crew on board.

Virgil 'Gus' Grissom commanded *Gemini 3*, thus becoming the first man to make two spaceflights. His copilot was John Young. It was a brief mission of just three orbits lasting a total of just under 5 hours on 23 March 1965. During that time they changed orbits, achieving an orbit that had a low point of just 53 miles (85 km). After splashdown Grissom was seasick. 'Gemini may be a good spacecraft but she's a lousy ship,' he said afterwards.

Gemini 4 astronauts Ed White
and Jim McDivitt simulate
launch procedures.

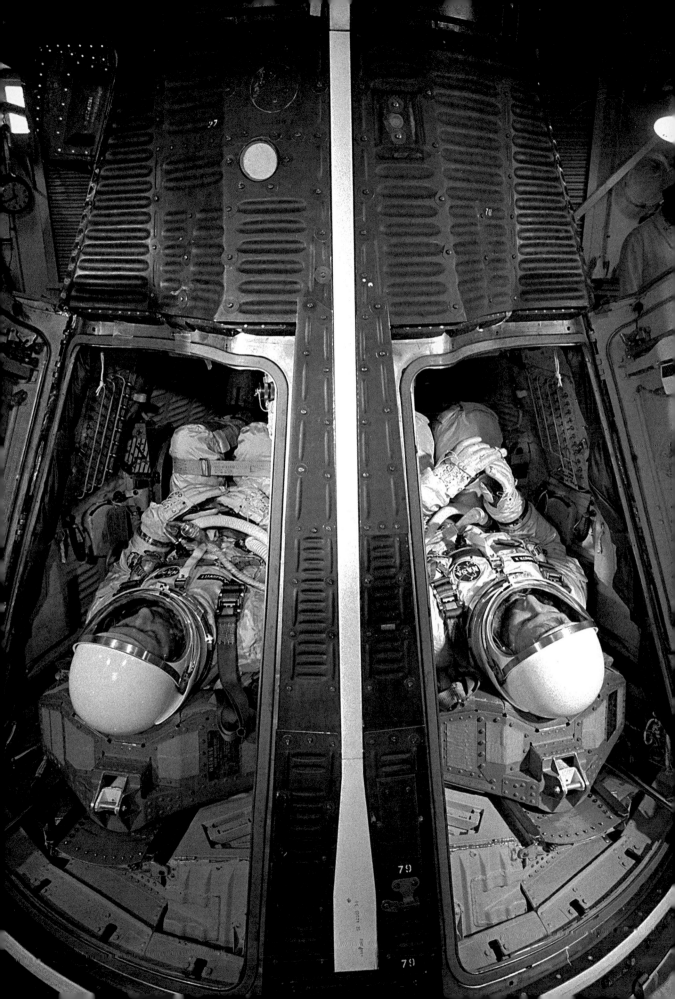

America's First Spacewalk

Looking back on the missions, for Gene Kranz, *Gemini 4* was one of the most exciting of the Gemini missions. It was his first as flight director:

> *We wanted to be the first to have an extravehicular operation; put a man out in space, free from the spacecraft. I got tagged to work with the team in building that EVA plan. And we were very imaginative; we called it Plan X. We'd finish our work here during the day; we'd go home, we'd eat, and then all the Plan X people would come back in and we'd work generally from about 6 or 7 in the evening until 1 or 2 in the morning, building the equipment, validating it in the altitude chamber, developing mission rules, etc.*

'I think if it hadn't been for Gemini, flying Apollo would've been nigh on impossible.'

ED WHITE

The USA's first spacewalk was carried out by Ed White of *Gemini 4*. The commander of *Gemini 4* was Jim McDivitt. He later commented that by the time the US started Gemini, that space race was over:

> *In Mercury, you couldn't manoeuvre. You could change its attitude but you couldn't change its flight path. Gemini you could. So, now you had to have the guy in the spacecraft working with the guy on the ground to know what was going on and where they were going, where they were, and what they were doing – what was going to happen. So, that worked out pretty well. As a matter of fact, I think if it hadn't been for Gemini, flying Apollo would've been nigh on impossible.*
>
> *My relationship with Ed couldn't have been better. He was the best friend I ever had. We lived, like I said, a block and a half or so apart. He was getting a Master's degree in aeronautical engineering, but he didn't have an aeronautical engineering undergraduate degree. So, we took a lot of classes together. We started flying together. I remember when the air force had its pre-NASA astronaut selection – I walked in the room in the Pentagon and Ed was already there and he says: 'I knew you'd be there!' And I said: 'I knew you'd be here, too!' Unfortunately as regards our EVA, we were beaten by the Russians. By what? A couple of weeks I guess. They were quiet up until a few days before the flight. I think originally it was to score the first!*

Gemini 4 was dispatched to space on 3 June 1965 – the US's first multiday mission – and once in orbit the crew turned their attention to the spacewalk.

> McDivitt: *When we got around to doing the EVA, Ed went to open up the hatch, but it wouldn't open. I said: 'Oh my God,' you know, 'it's not opening!' And so, we chatted about that for a*

Ed White floats in space during America's first spacewalk.
In White's right hand is a self-manoeuvring unit, which he
used to move himself around in the weightless environment.

minute or two. And I said: 'Well, I think I can get it closed if it won't close.' But I wasn't too sure
about it. I thought I could. But remember, then I would be pressurized. I wouldn't be in my sports
clothes, leaning over the top of the thing with a screwdriver as I had been in training. I'd be there
pressurized. In the dark. So anyway, we elected to go ahead and open it up.

White was outside for 21 minutes and had to be told to come back into the spacecraft by the Capcom
Gus Grissom. *Gemini 4* was headed for the Earth's shadow. 'This is the saddest moment of my life,'
replied White.

McDivitt: That was one of the reasons I was anxious to have him get back inside the spacecraft,
because I'd like to do this in the daylight, not in the dark. But by the time he got back in, it was
dark. So, when we went to close the hatch, it wouldn't close. It wouldn't lock. And so, in the dark I
was trying to fiddle around over on the side where I couldn't see anything, trying to get my glove
down in this little slot to push the gears together. And finally, we got that done and got it latched.

The Longest Flight

Two months later, on 21 August, *Gemini 5* was launched with Gordon Cooper and Charles 'Pete' Conrad on board. With the landing back on Earth taking place on 29 August, it was the longest space flight yet – the duration of the flight was equivalent to the time it would take to get to the Moon and back. This was made possible mainly by the first utilization of fuel cells; these generated extra electricity for longer flights.

Cooper: *Our Gemini 5 flight. We worked long and hard at it, and we couldn't do any EVA or do the other things because we were so loaded. We were absolutely crammed with equipment of all kinds they wanted us to have. We had the first fuel cell. We had the first radar. We had the first all up computer. These were all things that needed to be tested and proven.*

At the start of *Gemini 5* the oxygen pressure within the spacecraft dropped down to practically zero. According to the mission rules, the correct procedure if such an event occurred was to then switch everything off.

Cooper: *I had to go into total power down. So we powered everything down, brought everything down to low, low electrical power; and, of course, it happened again when we were out of radio range. So, as we came whistling over the horizon into communication, Houston realized we were all powered down and they really panicked for a moment; and it looked like we were going to have to re-enter another orbit later. But fortunately, and this is a story a lot of people don't know: When Pete and I were going through the altitude chamber with Gemini 5, we had to go through and do these various tests; and the tests finished on a Friday; the spacecraft was due to be shipped Saturday morning to the Cape from St Louis in order to stay on the time schedule. But one of the things we wanted to do was, we wanted to run some tests in the altitude chamber by decreasing both oxygen and hydrogen pressure, drastically, to see if the fuel cell would continue to run at altitude. NASA said: 'No, we can't afford the time. If we do it over this week, we can't afford to delay until next week to do it. And if we do it over the weekend, it would cost us triple time of overtime, so we're not going to do it.' So Pete and I went to Jim McDonnell, head of McDonnell Aircraft Corporation and told him the story on it, and he said: 'I'll pay for it. Let's do it.' So we spent the weekend in the altitude chamber at his cost doing the test; and if we had not done that test, we would have re-entered an orbit later.*

Flying in Formation

Three months later the US flew a double Gemini mission – *Gemini 6* and 7. Clearly the Gemini project was gathering momentum and confidence as the US surpassed the USSR in almost every aspect of manned spaceflight that mattered. Frank Borman flew *Gemini 7* with Jim Lovell on a record 14-day mission from 4 December to 18 December. According to Borman:

> *Gemini 7 was looked upon among the astronaut group as, you know, not much of a pilot's mission. Just sort of a medical experiment mission, which it was. Jim Lovell was a wonderful guy to spend 14 days with in a very small place. We had a lot of interesting things. You know, some of the doctors said: 'Oh well, in order to do that you're going to have to simulate it on Earth and see if you can stay in one G for 14 days.' And I, you know: 'They're out of their mind. Fourteen days sitting in a straight-up ejection seat on Earth? You're crazy!'*
>
> *We'd been up there for 11 or 12 days (I don't remember how long). And we were tired, and the systems on the spacecraft were failing. We were running out of fuel, and it was a real high point to see this bright light (it looked like a star) come up, and then eventually we could see it was a Gemini vehicle – Gemini 6. And we found that we could – we had very limited fuel – but we found that the autopilot for the controls were perfect. You could fly formation with no problem. And then Wally slapped up the sign: 'Beat Army'. Wally was always one to inject some levity into the programme. And, God bless him, he really did a good job in everything he did. He just has a different – he has that little quirk of being able to include some fun with things. I never had that. I didn't think much about the 'Beat Army' sign, although it was fun at the time. About the only thing that I really felt after two weeks like that were our leg muscles were shot. And it took about three or four days; and I guess you could feel it for a week or so afterwards. But it wasn't any big deal.*

In fact Wally Schirra and Tom Stafford nearly did not get their *Gemini 6* spacecraft into orbit. The plan was for *Gemini 6* to launch before Borman's flight and dock with an Agena target vehicle, but the Agena did not get into space and *Gemini 6* was only itself in space from 15 to 16 December. So a homing device was put on *Gemini 7* for *Gemini 6* to use.

'You could fly formation with no problem. And then Wally slapped up the sign: "Beat Army". Wally was always one to inject some levity into the programme. And, God bless him ...'

FRANK BORMAN

'They're in a roll and it won't stop'

EMERGENCY IN SPACE

KOROLEV'S DEATH AND THE LAST GEMINI MISSIONS
1965–1966

The period from 1965 to 1966 encompassed the end of some important phases in space exploration. It also marked the end of an era: the death of Sergei Korolev, who had been the driving force behind the Soviet space challenge. The period also saw the end of the controversial Russian Voskhod programme, as well as the completion of the more effective American Gemini missions.

Things were not going well in the USSR. As part of their unmanned reconnaissance of the Moon their Ye-6 automated lunar probe, designed to achieve the first soft-landing on the lunar surface, had failed time and time again. In reality it was not unusual. When the Americans tried to crash land a Ranger spacecraft on the Moon, taking pictures until the moment of impact, they failed many times before they succeeded. Between January 1963 and December 1965 there had been 11 consecutive failures for the Ye-6 programme, a record that had dampened the spirits of even the most optimistic Soviet engineers. Kamanin wrote in his diary: 'Korolev was more distressed by the setback than anyone. He looked dejected and appeared to have aged ten years.'

Death of a Soviet Hero
In fact, Korolev had been declining fast throughout 1965. In August he had complained about not feeling well because of abnormally low blood pressure, and in September he was afflicted by severe headaches. He also had a progressive hearing loss and a serious heart condition. He wrote to his wife:

> I am holding myself together using all the strength at my command ... I can't continue to work like this, you understand. I'm not going to continue working like this. I'm leaving!

He assessed what the Americans had achieved with their Gemini missions, knowing that even if the Soviets had flown their Voskhods they would have been unable to equal what the Gemini astronauts had done. In December Korolev underwent a series of medical tests in Moscow, which indicated he needed a minor operation to remove a bleeding polyp in his intestine. He spent his last day before the operation

A profile view of the Agena Docking Target Vehicle as seen from the *Gemini 8* spacecraft during rendezvous in space.

'Sergei Pavlovich is no longer with us. Now where did we leave off?'

VALENTIN GLUSHKO

at his office, staying late before being admitted to the Kremlin hospital the following day. He had already invited people to celebrate his 59th birthday at a party on 14 January.

Dr Boris Petrovsky, the USSR Minister of Health, removed a small polyp from Korolev's gastrointestinal tract, causing excessive bleeding. Petrovsky was an accomplished surgeon, but it seems he was unprepared for the complications that arose during the operation. Korolev had not told them that his jaw had been broken in prison from torture in 1938, which made it difficult for him to open his mouth wide. His unusually short neck compounded the problem, preventing the use of an intubation into his lungs. Instead, the surgeons performed a tracheotomy and inserted a tube in his neck. His jaw problem necessitated the use of a general anaesthetic despite his heart condition. Korolev bled profusely during the operation, and then Petrovsky found what he later described as an 'immovable malignant tumour which had grown into the rectum and the pelvic wall'. The size of the tumour, larger than a person's fist, was a shock to those in the operating room. Korolev was still bleeding profusely. Dr Vishnevsky, a cancer specialist, was called in and the two surgeons completed the operation, four hours after it had started. But half an hour later Korolev's heart stopped and they could not revive him.

His death shocked those involved in the Soviet space effort, but Korolev's arch enemy Glushko was apparently unperturbed. He was conducting a meeting when his Kremlin phone line rang. He heard the news, hung up and turned to his audience and said: 'Sergei Pavlovich is no longer with us.' He paused for a second and continued: 'Now where did we leave off?'

The End of the Voskhod Programme

Mishin was clearly the most likely choice as a successor to Korolev, having been groomed by him for almost a decade. His first job was to assess the state of the Voskhod programme. At the time of Korolev's death there were plans for three to four Voskhods and five Soyuz missions in 1966. The first one, *Voskhod 3*, was to be a long-duration mission with cosmonauts Volynov and Shonin. The spectacular success of the 14-day *Gemini 7* flight in December 1965 had given the Soviet mission even more of an impetus. The subsequent *Voskhod 4* would be a scientific flight, including artificial gravity experiments with test pilot Beregovoy and scientist Katys, while *Voskhod 5* would be a military mission that included cosmonaut Shatalov. As usual, *Voskhod 3* was timed to coincide with the opening of the 23rd Congress of the Communist Party in early March 1966. Dedicated to regaining the mission duration record claimed by *Gemini 7*, it was planned to be an outstanding publicity victory for the Soviet space programme. It never flew. No manned Soviet flights took place in 1966, while the US flew five Gemini missions, which tested all the techniques required for the Apollo project.

A Voskhod test spacecraft carrying two dogs was launched in February. The flight lasted nearly 22 days, but on their return the dogs were in a dreadful condition – wasting muscles, dehydration, calcium loss and problems in walking. Their motor systems did not return to normal until eight to ten days later. The United Press International Agency reported that the Soviet Union would launch a multicrewed spacecraft before the end of March 1966, in time for the 23rd Congress of the Communist Party.

However, long-duration ground tests of the life-support system did not go well. After 14 days, the Institute for Biomedical Problems had to abandon a test because of a deterioration in the cabin atmosphere. Parachute failures during recovery tests were common and worrying. Four cosmonauts were in training for the flight: Beregovoy, Shatalov, Shonin and Volynov, but as problems accumulated it became increasingly clear that there might never be a *Voskhod 3* mission. Soon it was cancelled.

Neil Armstrong's First Flight

There was no defining moment when Neil Armstrong decided that he wanted to become an astronaut. He had become a civilian test pilot, flying the advanced X-15 rocketplane. Apollo excited him and he applied to be an astronaut. His application papers arrived a week late, but fortunately a friend in the office put his papers in the relevant pile anyway. Deke Slayton called him in September 1962, and he became a member of the so-called 'New Nine'. Dave Scott came in the following year's selection.

They were launched in *Gemini 8* on 16 March 1966, following an Agena target vehicle into space. If *Gemini 7* was a routine mission, *Gemini 8* certainly was not. It was NASA's first serious space emergency. In orbit, Armstrong manoeuvred the spacecraft to make the first docking in space just a few hours after lift-off. Then the trouble began. Without warning they went into a dangerous spin.

Armstrong: *We first suspected that the Agena was the culprit. We had shut our own control system off, and we were on the dark side of the Earth, so we really didn't have any outside reference, or very good reference. I didn't actually notice when it started to deviate from the planned attitude. Dave first noticed it. Neither of us thought that Gemini might be the culprit, because you could easily hear the Gemini thrusters whenever they fired. They were out right in the nose, in the back. Every time one fired, it was just like a popgun, 'crack, crack, crack, crack'. And we weren't hearing anything, so we didn't think it was our spacecraft. Dave had the control panel for the Agena. That allowed Dave to send signals to the Agena control system. He was trying everything he knew, without success.*

When the rates became quite violent, I concluded that we couldn't continue, that we had to separate from the Agena. I was afraid we might lose consciousness, because our spin rate had gotten pretty high, and I wanted to make sure that we got away before that happened. Of course, once we separated and found out we couldn't … regain control in a normal manner, we recognized that it was a failure in our craft, not in the Agena. The reason we didn't hear it is, you only hear the thruster when it fires; you don't hear it when it's running steadily. I didn't know that at the time, but I figured it out.

1966

14 January Soviet space scientist Sergei Korolev dies. He is succeeded by Vasily Mishin

16 March *Gemini 8* is launched, with astronauts Neil Armstrong and Dave Scott on board, but a problem when docking with an orbiting *Agena* spacecraft means that an emergency re-entry manoeuvre has to be performed

3 June *Gemini 9* astronaut Gene Cernan tries unsuccessfully to become the first man to manoeuvre outside the spacecraft using a rocket pack

When Armstrong and Scott undocked from the Agena 'all hell really broke loose,' according to the flight director. The *Gemini*'s relatively slow but accelerating roll rate accelerated rapidly to the point where they were on the verge of losing consciousness. Neil Armstrong realized they had a jammed thruster. When they finally gained control they had used up so much fuel they had to perform an emergency re-entry.

Hunting the 'Angry Alligator' – *Gemini 9*

Gemini 9 was launched on 3 June 1966. Its crew was Tom Stafford and Gene Cernan. They had intended to dock with an Augmented Target Docking Adapter (ADTA) and perform a lengthy spacewalk, but the ADTA's protective covers had failed to come away, giving it an 'Angry Alligator' look.

As Tom Stafford was suiting-up for the flight, Deke Slayton, head of astronaut assignments, told the suit technicians to get out. He said: 'I need to talk to you, Tom.' He went on: 'Look, this is the first time we've got this long EVA, this rocket pack, and NASA management has decided that in case Cernan dies out there, you've got to bring him back, because we just can't afford to have a dead astronaut floating around in space.'

Stafford: *We'd never thought about that or anything. So I thought for a minute, and I said: 'Jesus Christ, Deke.' I said: 'Look, to bring him back, I've got to have the hatch open because of that cord going out.' And he says: 'Well, what do you want me to tell NASA management?' I said: 'You tell them that when the bolts blow, I'm the commander and I'll make the decision. That's it.' 'All right.' He left. Gene says: 'Hey, Tom, Dick was in there talking to you quite a while. What did he say?' I said: 'He said he just hoped we'd have a good flight.'*

So we got all ready to go and we launched, and it lifted off in June. I remember coming up to it, and you could see the constellation Antares. There was a full moon out. We got up close, I could see this weird thing. I came right up close to it, and it just broke out in sunrise, and here was the shroud, like that. I call it 'The Angry Alligator'. Then came time for Cernan to go EVA, and they wanted him to go out and cut loose the shroud, to cut it loose. I looked at it. I could see those sharp edges. We had never practised that. I knew that they had those 300-pound springs there, didn't know what else. So I vetoed it right there. I said: 'No way.'

Cernan goes out, and the first thing he does is place the rear-view mirror on the docking bar. He's huffing and puffing. He's torquing the hell out of this spacecraft, and I'm pulsing it back to be sure none of the thrusters fire on him. He goes out in front and does a few little manoeuvres

and he's having a very difficult time. There was nothing for him to hold on to. Remember, the Gemini suit was a very easy suit to wear. It was lightweight, twenty-five, twenty-eight pounds, and when it was uninflated, it was limp. So he's hanging onto the back end of the Gemini, going through this check-out procedure for the rocket pack. Then he says: 'Tom, my back's killing me. It's burning up. It's really killing me.' I says: 'What?' He says: 'My back.' I could look in the rear-view mirror, and I could see the Sun. Of course, you never look directly at it. I said: 'Do you want to get out of the spacecraft?' He said: 'No. Keep going, but my back is killing me. It's burning up.' So he finally, just before sunset, gets turned around into the seat. We had two lights back there. One of them burned out for some reason, during the vibration of the launch anyway. We had one light. And then a couple of minutes after sunset, he was strapping himself in. I was down to a couple of steps. You know, at sunrise I would cut him loose. He fogged over. Whop! He could not see. It was just like that, fog. So we did defog on the visor, and he had overpowered that little water evaporator so much, it was unbelievable.

And then we started to lose one way of two-way com. It was real scratchy. So he could hear me. So I worked out a binary system. I said: 'Look, if you can hear me, make a noise for a yes. For a no, make two noises.' So I'd hear a 'squawk', and for 'no', I'd hear a 'squawk, squawk'. So I said: 'Can you see?' I'd hear a 'squawk, squawk.'

> ## 'We just can't afford to have a dead astronaut floating around in space.'
>
> DEKE SLAYTON

I think we went south of Hawaii, then, before we hit the West Coast of the US. We went a long time. It was night time. We saw the Southern Cross go by. What a hell of a lonely place this is. Here you're 165 miles up, you know, flying, pressurized. Your buddy's 25 feet back there. He can't see, and we'd lost one way of two-way com. There's not a thing you can do until you get daylight. So it came up daylight. He could see it was daylight. I said: 'Okay, Gene. If this doesn't burn off fast, we're going to call it quits and get out of there.' So after five or ten minutes, nothing happened. So I said: 'Okay. Call it quits, Gene. Get out of there.' He couldn't see. He was absolutely blind and 145 feet away.

We hit the West Coast. I said: 'Look, I've called it off. He's fogged over, he can't see. We've semi-lost one way of two-way com. I'm not going to fly the rocket pack.' My main thing is to get him in before the next sunset. So we got all squared away, and he got in, and we worked out this manoeuvre. Still, he couldn't see hardly a thing. So he got down. I helped guide him down. He knew how to feel on this thing. So he came in closer, and I just reached over and grabbed this over-the-centre mechanism and slammed it. Then as the pressure came up, our suits decreased. So finally he got back in his seat, raised his visor, and his face was pink, like he'd been in a sauna. He says: 'Help me get off my gloves, too.' So I helped him get his gloves off, and his hands were

The ADTA target vehicle for *Gemini 9*, with its protective covers still attached, giving it an 'Angry Alligator' look.

absolutely pink. So I took the water gun and just hosed him down. You shouldn't squirt water around in a spacecraft. Turns out he lost about ten and a half pounds in two hours and ten minutes outside. That was the third day. We landed the next day. By the way, the ride on the carrier, I want to tell you about that. We got the suit back to Houston, and the next morning they still poured a pound and a half of water out of each boot. Finally, after we did splash down, after the final thing, we're back in the crew quarters having a drink, I told him what Deke said.

The Last Gemini Missions

The next mission, *Gemini 10*, launched on 18 July with its crew of John Young and Michael Collins, also had its problems. They also could not see during the spacewalk towards the *Agena*.

Collins: *The method that we used on Gemini 10 to purge the system, to absorb the exhaled carbon dioxide from your body, were canisters of lithium hydroxide. The stale air went through the lithium hydroxide and it came out purified. Lithium hydroxide is a granular sort of material, and our best guess was that somehow lithium hydroxide had escaped from some canister and had gotten into the nooks and crannies of the system in the pipes and that there was some triggering mechanism having to do with depressurizing the spacecraft that caused that lithium hydroxide to start billowing up. It went through, and it can be an irritant, and that's what it was. But to the best of my knowledge, they never established that beyond the shadow of a doubt. All I know is that I couldn't see and John couldn't either, and it was frightening for a moment, because the hatch on*

Gemini was not a very straightforward thing. In other words, you just didn't go 'clunk, latch'. You had to look up, and there were little levers and whichnots that had to be fiddled with, and then you had to make sure that all your hoses and stuff were not going to get in the way, and then you had to come down in a certain way and you had to get your body underneath, your knees underneath the instrument panel and kind of ratchet your body down, and it was a tight fit. So it was the kind of stuff that, with practice, you found it became easy to do, but ... it wasn't something you ever had trained to do or thought you would have to be doing by feel alone.

By the time we got to Gemini 12 with Buzz Aldrin there were handholds and work stations. On the Shuttle, you see it in space. I mean, they don't go out without being anchored in two or three different ways. But we were stupid; we hadn't thought of that. So, the point is, I was going over to this Agena, propelling myself with this dorky little gas gun. So anyhow, I was using this little gun to get over to something, to grab something that had not been designed to be grabbed, and I'm in this bulky suit that I described before, it doesn't want to bend too well, I'm immobilized. I'm having a tough time as I'm going along, pitch, rolling and yawing, trying to keep this dorky little gun through the centre of mass of my body, and then I arrive at this goddamned Agena, which is not meant to be grabbed, and I've got to grab it. So, the first time I grabbed it, I went to the end of it, and it had a docking collar. Docking collars are built to be nice and smooth so that the probe that goes into them will be forced into them. They have smooth lips and edges on them, and that's what I was grabbing. Well, I grabbed the docking collar. It wasn't meant to be grabbed, bulky glove, and my momentum is still carrying me along, so I just slipped, and as I went by, then I went cartwheeling ass over teakettle, up and around and about, until I came to the end of my tether, I went back to the cockpit, and then John Young got a little bit closer to the Agena the second time, and when I went over to it the second time I was able to get my hand down inside a recess between the main body of the Agena and the docking collar where there were some wires, and grab some wires.

There were a couple of rendezvous on Gemini 10. We rendezvoused with two different Agenas: our own Agena, call it Agena 10, and then a dead, inert Agena that had been used by the Gemini 8 flight, that had been up in the sky for a couple of months just sitting there. These Agenas were different in two respects. Agena 10 we could ask questions and it would answer. It had a transponder. So we could ask it: 'How far away are you?' and it would tell us. The Gemini 8 Agena had dead batteries. Its transponder could not reply. So when we asked it questions, it would not answer. This meant that we could not find out how far away from the Agena we were. We had to just deduce our range by the apparent size of the Agena or the actual size.

The Gemini project closed in November 1966 after two more missions. It had accumulated 80 man-days in space over ten missions. It had performed orbit changes, spacewalks, rendezvous and re-boosts. It had seen adversity overcome in orbit. All those techniques would be needed for the Apollo project. The way to the Moon was open. But 1967 was to be a very bad year for everyone.

'We're on fire! Get us out of here'

DEATH AND THE ASTRONAUT

APOLLO 1 AND SOYUZ
1967–1968

Despite failures as well as successes, both the Russian Voskhod and the American Gemini programmes seemed to be paving the way towards the next prize in the space race – landing a human safely on the Moon. But the dangers and technical difficulties inherent in achieving this glittering prize were made starkly clear to the world in a series of catastrophes that claimed the lives of astronauts, cosmonauts and others.

The Apollo programme was designed to land humans on the Moon and then bring them safely back to Earth – a goal announced by US President John F. Kennedy in 1961. While often cited as one of the greatest achievements in human history, it started with disaster.

The *Apollo 1* Disaster

On 27 January 1967 at launch pad 34 at Cape Canaveral, three astronauts were performing a so-called 'plugs-out' test, in which the newly designed Apollo capsule positioned above a *Saturn 1B* rocket would be isolated from external equipment. The crew of Gus Grissom, Ed White and Roger Chaffee had been chosen for the first manned Apollo flight. They were sealed inside the capsule, breathing a high-pressure 100 percent oxygen atmosphere. The crew members were in their horizontal couches running through a checklist when a voltage spike was recorded at 6.30 and 54 seconds. Ten seconds later Chaffee said: 'Hey … ,' and scuffling sounds were heard. Grissom shouted: 'Fire,' followed by Chaffee, who said: 'We've got a fire in the cockpit.' Then White repeated: 'Fire in the cockpit.' Seconds later Chaffee yelled:

We've got a bad fire! Let's get out! We're burning up! We're on fire! Get us out of here!

The crew in training for the ill-fated *Apollo/Saturn 204* mission, later to be renamed *Apollo 1*.

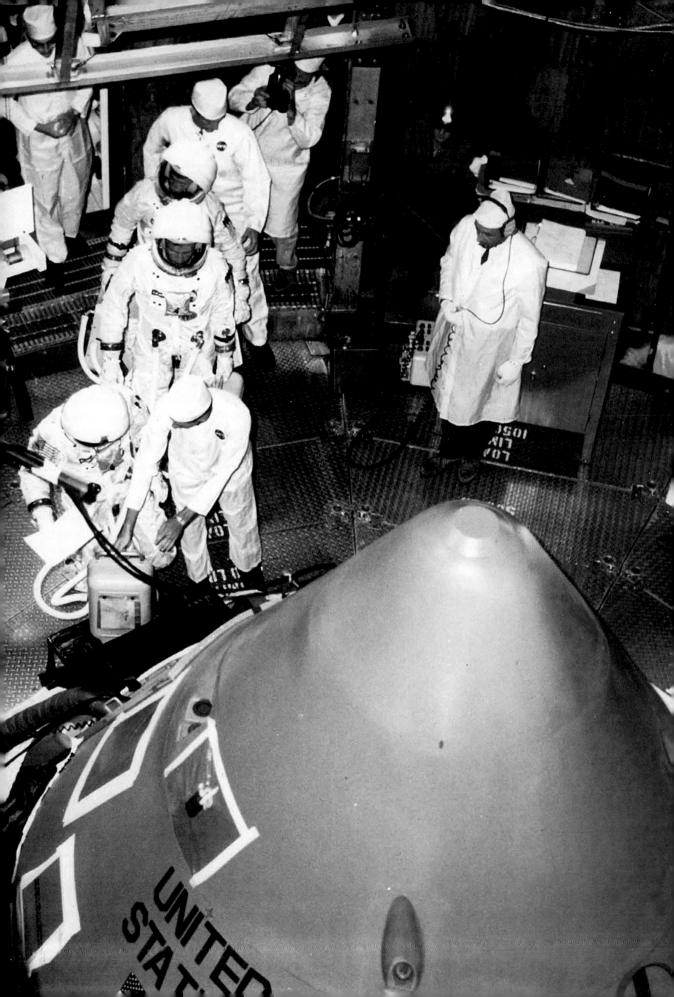

Just 17 seconds after the first indications of fire, the transmission ended with a scream as the capsule ruptured due to the expanding gases. Toxic smoke was leaking from it. The astronauts had tried to open the hatch but it was too awkward, too complicated. The pure oxygen atmosphere did not give them a chance. Post mortems showed they all had extensive third-degree burns and had died due to a combination of smoke inhalation and burning. To this day no one knows what caused the initial spark. The subsequent inquiry found that the documentation was so poor that no one was even sure what was within the spacecraft at the time of the accident.

> Borman: *We were having dinner with some friends on a lake in Huntsville, Texas; and a highway patrolman knocked on the door and said that I was supposed to call Houston right away. Susan and I left and drove back to Houston and went over to Ed White's house, because Susan was close to Pat White.*

> Armstrong: *I was in Washington. The president was signing the Outer Space Treaty with other nations that kept the Moon as the property of all people. It was a non-staking-a-claim treaty. I'd known Gus Grissom for a long time. Ed White and I bought some property together and split it. I built my house on one-half of it, and he built his house on the other. We were good friends, neighbours. I suppose you're much more likely to accept loss of a friend in flight, but it really hurt to lose them in a ground test. That was an indictment of ourselves. I mean, it happened because we didn't do the right thing somehow.*

When Gene Kranz came off Gemini and turned his attention towards Apollo he was in for an unpleasant surprise:

> *I was really shocked by how far we had yet to go before we could pull together a coherent Apollo operation with the same quality that we were now experiencing in the Gemini operation. And this was particularly true in our relationships with North American Rockwell who made the Apollo Command Module. Rockwell is a very good contractor, but they hadn't been flying in space before. All of our experience had been with McDonnell. And Rockwell was used to building fighter airplanes, rolling them out of the factories, etc., and they weren't about to listen to anybody that wasn't a test pilot. This friction in January, I think, led to the disaster that we had with the pad fire. The fact is that we really weren't ready to do the job, and yet we were moving on. We were sitting there that*

The prime *Apollo 1* crew (from left) 'Gus' Grissom, Ed White and Roger Chaffee, who all died in a fire during testing.

day, running the test. I had done the shift prior to the fire, and things weren't right that day, and I knew they weren't right. And yet I continued on. I think everybody that was working that test knew things weren't right. We weren't ready! But nobody stood up and assumed the accountability and said: 'We're not ready. It's time to regroup.' And I think this was one of the very tough lessons that came out of Apollo 1, that we said, 'From now on, we are forever accountable for what we do or what we fail to do.'

We did have an onboard tape that was probably running. It was probably burnt up. But, where they had the air-to-ground, which was from the spacecraft to the blockhouse, and in the 21 seconds from the time you first heard the noise to the time it was over, no one was exactly sure what they said. So, I remember the very next day (I think it was the next day) I was down at the Cape. We flew back either – this was Friday – we flew back either on Saturday or Sunday. I flew back once, I think, to take Gus's uniform down there for the burial. And then I flew back and we stayed there – Donn and I stayed there – and we sat down with the tapes. We had to – I think Frank Borman and Donn Eisele and I, because we knew the guys, we knew their voices, we sat down and went through this. And even then, we couldn't agree exactly on what went on. And they wanted to get it down to timing. So, I ended up taking those tapes up to Bell Labs up in New Jersey, where they broke it down, did all the magic things they do with it. You know, today it would've been easy with digitizing, but it was tough in those days and they still had some controversy afterwards.

McDivitt: *We were doing those same things in Gemini and Mercury. We could've had exactly the same problem with Gemini and Mercury. We were pressurizing the spacecraft at 5 psi over atmospheric, which was 20 psi. We had a 100 percent oxygen environment. I did the same test on top of the Gemini that they were doing at the time that the fire occurred. And we did it on every Gemini spacecraft. I think we did it on every Mercury spacecraft, too. To this day, nobody knows how the fire started. But we just had a lot of bad circumstances come together. And some of the North American people maintain to this day that they were never told that the spacecraft would ever be tested in this configuration. If they didn't know it, they were the only people in the whole world that didn't know it. But, you know, everybody had their own idea how this was going to work, I guess. But it was one of those circumstances. You had all this flammable material in there and a 100 percent oxygen environment at 20 psi. That's seven times more oxygen than we have in this room right now.*

Schirra: *I was annoyed at the way what became Apollo 1 came out of the plant at North American Aviation's plant in Downey, California. It was not finished. So it was shipped to the Cape with a bunch of spare parts and things to finish it out. And that, of course, caused this whole atmosphere of developing where I would almost call it a first case of bad 'go' fever. 'Go' fever meaning that we've got to keep going, got to keep going, got to keep going! When my crew did the test that was followed by Gus and his guys, we were in sea-level atmosphere; no pure*

'If I can't talk to the blockhouse, how the hell are we going to go to the Moon with this damn thing?'

GUS GRISSOM

oxygen. We were in shirtsleeves. And there were things going on I didn't like at all. I was no longer annoyed; I was really pretty goddamn mad! There were glitches, electronic things that just didn't come out right. That evening I debriefed with Apollo Spacecraft Programme Office Manager Joe Shea and Gus. And I said: 'If there are any things that go wrong, like a glitch in the electronic circuits and bad sounds, scrub!' Because Gus and his guys were going to do it in pure oxygen and in an environment that's not very forgiving. We didn't realize how unforgiving it was at that point. We'd gone through the same environment with Mercury and Gemini and made it through. Not that I think of it in that way, but that's how I look at it in retrospect. Gus, I can recall saying: 'If I can't talk to the blockhouse, how the hell are we going to go to the Moon with this damn thing?' That's how bad the communications were. He should have scrubbed. He didn't. He was himself involved in 'go' fever.

After the accident Frank Borman was asked if he thought that NASA could not recover from the disaster. He replied: 'Never, not for one instant.'

Russia Aims for the Moon

Despite later denials, the Soviets were desperate to beat the Americans to the Moon – too desperate. They had three manned programmes on the go. The first was a manned flight around the Moon called the L1 project. The second, the L3 project, was a manned landing that required a lunar lander and a giant booster called the *N1*. The third was their orbital missions. For all three of these manned programmes, the Soyuz spacecraft was to be the centrepiece.

The first mission planned was intended to be a 'spectacular' – the docking of two Soyuz spacecraft in Earth orbit, followed by the transfer of two crew members from one vehicle to the other via a spacewalk. Soviet space leaders believed that this single mission would overshadow the achievements of Gemini. Since September 1965, four air force cosmonauts had been training for the commander's spot on the two Soyuz spacecraft: veterans Bykovsky, Gagarin, Komarov and Nikolayev. Vladimir Komarov was the leading contender for the commander aboard the active Soyuz.

Engineers began the ground testing of the first flight model of the Soyuz spacecraft on 12 May 1966, and it had echoes of the *Apollo 1* Command Module. There were many problems – over 200 known faults. Instead of the anticipated 30 days, it took four months to debug the ship and even then the cosmonauts had no confidence in it. One night Komarov confided to a colleague:

I'm not going to make it back from this flight. If I don't make this flight, they will send the backup pilot instead. That's Yuri, and he'll die instead of me. We've got to take care of him.

There are some reports that Gagarin tried to get Komarov removed from the flight, knowing that it would then have to be cancelled because they would not risk him on such a mission.

Fire on the Soyuz Launchpad

There were severe problems with the Soyuz's parachute system. Two of the seven drop tests from an aircraft failed. Kamanin wrote in his diaries:

One has to admit that the Soyuz parachute system is worse than the parachute system of the Vostoks and the spacecraft isn't much to look at in general: the hatch is small. The communications equipment is outdated, the emergency rescue system is primitive and so on. If the automatic docking device turns out to be unreliable (which cannot be ruled out) our space programme will be headed for an ignominious failure.

The first Soyuz test spacecraft lifted off successfully from Baikonur on 28 November 1966, entering a lower orbit than expected. The Soviet news agency TASS designated the spacecraft *Kosmos-133* and did not indicate that the flight had any connection with the manned space programme. The mission ran into problems almost immediately, making the spacecraft unusable. The launch of the second test Soyuz, with which it was to have automatically docked, was cancelled.

Two weeks later the Soviets tried again. As the rocket ignited they noticed that it did not seem to be working properly. Then it shut down, remaining on the launch pad. Steam filled the area as thousands of gallons of water poured onto the launch mount. About 27 minutes after the abort, the launch escape system suddenly ignited. In seconds the rocket's third stage caught fire. Kamanin described the scene:

I ran to the cosmonauts' house and ordered everyone who was there to quickly go from the rooms into the corridors. It proved to be a timely measure: within seconds a series of deafening explosions rocked the walls of the building, which was located 700 metres (2296 ft) from the pad. Stucco fell down and all the windows were smashed. The rooms were littered with broken glass and pieces of stucco. Fragments of glass hit the walls like bullets. Clearly, if we had remained in the rooms a few seconds longer we would all have been mowed down by broken glass. Looking out through the window openings I saw huge pillars of black smoke and the frame of the rocket devoured by fire.

The next Soyuz test spacecraft was prepared for its two-day mission in early 1967. It took off on 7 February, and this time reached orbit successfully. TASS announced the flight as *Kosmos-140*, another in a long series of nondescript generic satellites with no stated mission. Trouble started on its fourth orbit. The solar panels were not being directed towards the Sun, and so the craft's batteries could not be charged.

Vladimir Komarov, seen here training in a simulated spacecraft, died when the Soyuz 1 craft he was piloting crashed during landing.

In addition, fuel levels in the manoeuvring thrusters were down to 50 percent remaining. All was not forlorn, however. It was thought that all of these problems could have been overcome if a cosmonaut had been on board. The remaining systems, such as life-support, the main engine and thermal control, worked well. Nevertheless, the cosmonauts could see that the Soyuz capsule was a reckless gamble. Despite these problems, officials kept pointing out that a cosmonaut had not been in space for nearly two years.

A Death in *Soyuz 1*

The State Commission decided to press ahead with the dual manned launches, setting 23 April as the launch date. The cosmonaut in the active *Soyuz 1* would be Komarov. The following day, as the spacecraft was flying over Baikonur, the passive *Soyuz 2* would be launched with Bykovsky, Yeliseyev and Khrunov on board. After docking, Yeliseyev and Khrunov would spacewalk to the *Soyuz 1* which, with a crew of three, would return the following day. *Soyuz 2*, with a crew of one, would also return that same day.

A film exists of Komarov being driven to the launch pad in a bus. He looks condemned. Kamanin and Gagarin accompanied him to the rocket. Gagarin went with him all the way to the top of the rocket and remained there until the hatch was sealed. *Soyuz 1* lifted off on time. Komarov was the first cosmonaut to make a second flight. He was 40 years old. He ran into problems straight away.

One of the two solar panels did not deploy, resulting in a shortage of power for the spacecraft's systems. Komarov also had problems orienting *Soyuz 1*. By orbit 13 the stabilization systems had failed completely. Unconfirmed reports suggest that Komarov even tried to knock the side of the space ship to jar open the stuck panel. Due to dwindling power, he could not stay in space for long. The second Soyuz flight was cancelled and plans made for an emergency re-entry. The three cosmonauts of *Soyuz 2* pleaded to be allowed to launch, arguing that they could perhaps perform a spacewalk to free the jammed solar panel on *Soyuz 1*. But they were turned down.

Chertok carefully checked over the set of instructions that Gagarin personally transmitted to Komarov. In the final seconds before loss of contact, Mishin and Kamanin both wished Komarov good luck. At the appointed time the re-entry rocket did not fire. Communication with Komarov was re-established and the problem rectified. Another attempt was made. Komarov did not have many more chances left. Miraculously,

Komarov manually oriented *Soyuz 1* and managed to fire the rocket to slow the capsule down so that it could re-enter the atmosphere – the so-called de-orbit burn. About 15 minutes after retrofire, there was the expected break in communications as Komarov's capsule entered an ionization layer. A few minutes later, Komarov's voice cut through the radio silence sounding 'calm, unhurried, without any nervousness'.

The pilot of one of the search-and-rescue helicopters flying east of Orsk reported that he could see the Soyuz capsule. When they reached the landing site it was clear that there had been a disaster. The re-entry capsule was lying on its side, and the parachute could be seen alongside. The capsule was surrounded by clouds of black smoke. It was crushed, on fire and completely destroyed. The parachutes had not worked. Komarov's body was found in the crushed capsule. Medics gathered what they could and a few days later his ashes, like those of Korolev, were interred in the Kremlin Wall after a state funeral. But so great was the destruction of the capsule, that more remains were found later and buried at the crash site.

Chasing a Space 'Spectacular'

To add to the problems with the recovery from the Soyuz disaster, Premier Brezhnev wanted a space 'spectacular' to coincide with the 50th anniversary of the Great October Revolution due in November 1967 – preferably a cosmonaut trip around the Moon. Indeed, just days after Komarov's death, chief designer Mishin set out a new plan for the circumlunar project that envisaged four automated spacecraft flying around the Moon between June and August 1967, followed by three piloted flights occurring in time to make the November 1967 political deadline.

Many in the West believed that the Moon was the Soviets' goal. In May 1967 Gemini astronauts Michael Collins and David Scott visited the Paris Air Show at the same time as Soviet cosmonauts Pavel Belyayev and Konstantin Feoktistov. It was only a month after Komarov's death, and the Americans gave their condolences. What they were told was that there would be several Earth orbital test flights that year followed by a circumlunar flight. Collins later recalled that Belyayev said he expected to make a circumlunar flight in the not-too-distant future. Later that year academician Obraztsov said: 'The very next milestone in the conquest of space will be the manned circumnavigation of the Moon, and then a lunar landing.'

It was clear that the Soyuz docking and spacewalk mission – practice for when the crew transferred to the Lunar Transit Craft (the L1) – would be delayed, and so the plan was changed. There would be no transfer. Instead, cosmonauts would launch in the L1 direct, which would be carried on a more powerful rocket – the *Proton*. Because of this, two additional automated circumlunar missions were added to the flight sequence, making a total of six robotic flights before a piloted one. If they all went well, there was a chance they could fly a man around the Moon by the November 1967 political deadline.

Once again it was a highly dangerous strategy. Testing on the spacecraft had hardly started and was beset with poor standards by contractors. Communist Party and government leaders were anxious, knowing that the first launch of the Americans' mighty *Saturn V* rocket was due for late 1967, while their own giant *N1* booster – the key to their manned landing – was still many months away from launch. In August 1967 Secretary of the Central Committee for Defence and Space Ustinov was infuriated. He told Mishin: 'We have a celebration in two months, and the Americans are going to launch again, but what about us? What

'All of us need a successful launch like a breath of fresh air.'

NIKOLAI KAMANIN

have we done?' As the summer wore on some degree of sense prevailed. It was realized that the November deadline was impossible. With the pressure off for a little while, Mishin could think instead about beating the Americans to a flight around the Moon. Following the Apollo fire, US manned flights would not take place until mid-1968 at the earliest. It was a breathing space.

The *Saturn V* Test

Something important did take place in the month of the 50th anniversary of the Bolshevik Revolution, however. That was the launch of *Apollo 4* – the first test, unmanned, of the *Saturn V* rocket that would take Americans to the Moon. Eighty-nine trucks of liquid oxygen, 28 trucks of liquid hydrogen and 27 of RP-1 (refined kerosene) were delivered to Cape Kennedy to fuel this huge machine. Ignition of the rocket's first stage saw five mighty F-1 engines burning RP-1 and liquid hydrogen, and providing 7.5 million pounds of thrust. The second stage brought in four powerful J-2 engines burning liquid oxygen and hydrogen. The third-stage engine, a single J-2, also worked flawlessly. Von Braun watched it rise into the sky. He had fulfilled his promise.

Soviet Moonship Plans Collapse

In January 1968 the L1 cosmonauts began training in a specially built simulator, carrying out at least 70 trial runs. The following month Mishin and Kamanin agreed on the selection of four crew commanders to train for the first few missions: cosmonauts Bykovsky, Leonov, Popovich and Voloshin. They, along with eight others, were engaged in an intensive programme throughout 1967–8, but it seems that they had little confidence in the spacecraft. Kamanin recalled in early March that the cosmonauts were working diligently and knew the craft well:

> Perhaps, it is precisely because the cosmonauts excellently know all the strong and weak points of the craft and the carrier rocket that they no longer have their initial faith in the space hardware.

There was no lunar launch window in early 1968, so Mishin and Chelomei agreed to launch a test spacecraft out into deep space to a distance of about 205,100 miles (330,000 km) – about the distance to the Moon – and bring it back to Earth, thus simulating an actual circumlunar flight. But there was a feeling of inevitable failure that was difficult to shake off. Kamanin echoes it in his diary:

> All of us need a successful launch like a breath of fresh air. Another failure would bring innumerable troubles and may kill the people's confidence in themselves and the reliability of our space equipment.

On 2 March 1968 the test moonship lifted off into a circular Earth orbit. Soon afterwards the Block D booster stage fired for 459 seconds to place the craft into an elliptical orbit, with a high point of 220,000 miles (354,000 km) – lunar distance. TASS did not announce anything of note about the launch, except to name the spaceship *Zond 4* ('*Zond*' being the generic Russian word for 'probe'). At the end of its mission the spacecraft evidently passed through the atmosphere safely and was about to deploy its parachutes near the West African coast when the emergency destruct system on the descent capsule ignited. Such a destructive charge had been incorporated into the spacecraft for fear that it might fall into Americans hands. Another test flight took place the following month. This time the third stage failed to ignite and the emergency rescue system was activated. The political leadership was extremely worried by the accumulating series of failures in the programme. Despite them, Mishin was ordered to accelerate the pace of work on the L1 in order to launch a crew around the Moon by October 1968.

A test flight around the Moon was scheduled for July, but days before the launch, while the rocket and spacecraft were being tested on the pad, the Block D stage exploded, killing one person. The aftermath of the accident was extremely dangerous, too, and observers watched in terror. The lunar spacecraft and part of Block D had become unstable, threatening to fall and explode at any time. Engineers risked their lives dismantling the explosive wreckage. The July and August lunar launch windows were abandoned.

1967

27 January Three US astronauts die in the fire inside an Apollo spacecraft during on-pad tests

24 April Cosmonaut Vladimir Komarov dies on landing after *Soyuz 1* test flight

9 November The first *Saturn V* rocket blasts off, carrying an unmanned *Apollo 4* spacecraft

1968

2 March Russian *Proton* rocket launches the prototype of the L1 circumlunar spacecraft, designated *Zond 4*

14 July An explosion at the Proton launch complex with L1 (Zond) spacecraft in pre-launch processing kills one person, delaying the programme

18 September *Zond 5*, the prototype of the Russian L1 spacecraft for manned circumlunar flight, circles the Moon

Soyuz Dock in Orbit

Meanwhile the manned Soyuz flights in Earth orbit were continuing. Two automatic Soyuz craft were prepared in order to practise rendezvous and docking manoeuvres. For the first time in the Soyuz programme, all of its systems were working without fault when it reached orbit; the solar panels deployed as intended and the Igla ('needle') radio docking system was functioning correctly. On the second day of the flight there were some minor problems, but the State Commission nevertheless gave the go-ahead for the second Soyuz launch. The craft was launched on 30 October. Within 62 minutes of its launch both vehicles were docked – the first automated docking ever carried out. After the vehicles were linked, however, ground controllers discovered that there had not been a 'hard' docking, because there was a gap between the two ships. Upon analysis this was considered to be a minor problem, and after three and a half hours the vehicles separated.

Russia's 'Big' Rocket

By March 1968, NASA had still to recover from the *Apollo 1* tragedy and was still months away from flying a piloted Apollo spacecraft in Earth orbit, let alone in lunar orbit. Many Soviet officials believed that it would take a miracle to successfully carry out a sequential series of successful Apollo missions in the months leading to a first landing by the decade's end. But in many ways, the Soviets were viewing American capabilities like their own. In a diary entry in March 1968, Kamanin wrote:

> It took us three extra years to build the N1 and the L3, which let the United States take the lead. The Americans have already carried out the first test flight of a lunar spacecraft, and in 1969 they plan to perform five manned flights under the Apollo programme.

In the summer of 1968 the US press was full of rumours about the impending launch of a super-heavyweight Soviet rocket comparable to the *Saturn V*. The head of NASA, James Webb, said:

> There are no signs that the Soviets are cutting back as we are. New test and launch facilities are steadily added, and a number of spaceflight systems more advanced than any heretofore used are nearing completion.

Later, George Mueller, the NASA Associate Administrator for Manned Space Flight, told Apollo contractors that the Soviets were developing a 'large booster, larger by a factor of two, than our *Saturn V*'.

Those responsible for this large booster, the *N1*, knew it was a gamble. The so-called Council for the Problems of Mastering the Moon met on 9 October to discuss the status of the Soviet lunar landing programme. Mishin said that the first *N1* flight model would only be able to lift 76 tons, but a modification of the second stage would mean that the 95 tons needed for a lunar landing by a single cosmonaut could be lifted. More improvements might make it possible to take two cosmonauts to the Moon's surface. Academy of Sciences president Keldysh was one of the strongest supporters of the two-cosmonaut plan, considering sending one cosmonaut as very risky. But then he made the reckless proposal that they should consider landing two cosmonauts on the Moon on the very first launch of the *N1*! If that was impossible, then the mission should attempt to land a lone cosmonaut. Brezhnev is reported to have said:

> We should prepare for a manned mission to the Moon straight after the first successful launch of the N1, without waiting for it to be finally developed.

Brezhnev's demands emphasize the gap between the people building the spacecraft and those who controlled the finances. One could say that the politicians did not understand the true engineering facts of the situation, but then again, the engineers themselves seemed to be turning a blind eye towards them.

Earthlings Circle the Moon

As summer gave way to autumn, the Soviet piloted circumlunar programme was getting into deeper trouble. In four tests since late 1967, there had been three complete failures and one partial success – the mission of *Zond 4* in March 1968. And another L1 spacecraft had been destroyed during ground preparations for a test launch scheduled to take place in July 1968.

It was under this cloud that the first of the three remaining L1 spacecraft arrived at the Baikonur Cosmodrome for a new series of attempts, the first coinciding with the lunar launch window in September. This time the L1 launch was perfect. The *Proton* booster lifted off on 15 September with the Moon visible tantalizingly above the launch pad. At an altitude of 100 miles (160 km), the third stage ignited, and after a tense 251 seconds the rocket went into a perfect Earth orbit of 119 by 136 miles (191 by 219 km). After a circuit around Earth the Block D fired a second time to send it towards the Moon. Shortly afterwards the Soviet press announced the launch, designating the mission *Zond 5*. It was the first time in their circumlunar programme that a spacecraft had been successfully sent towards the Moon. Engineers and cosmonauts were jubilant. A few days later it circled around the Moon at a distance of 1218 miles (1960 km) and was flung onto a return trajectory towards Earth. It splashed down in the Indian Ocean on 21

Soviet *N1* rockets on the launch pad at Tyura-Tam in July 1969. The *N1* was designed for the Soviet human lunar missions programme.

September and was hauled in by the Soviet navy, which in turn was watched closely by the US navy. The tortoises and other animals onboard survived their ordeal – the first earthlings to go round the Moon.

The *Zond 5* mission was the first real success in the L1 Moon programme. It allowed the USSR to make plans for flying a crew on a circumlunar mission in January 1969, dependent upon two more successful L1 flights. The cosmonauts had almost completed their training programme, and it was hoped that one of the crews would make history as the first humans to fly from Earth to the Moon. But the men training for a circumlunar mission were not the only cosmonauts preparing for spaceflight in the fall of 1968. By August, cosmonauts Beregovoy, Volynov and Shatalov had completed their preparation for the first piloted Soyuz mission since the *Soyuz 1* tragedy more than a year before.

'That was a real kick in the pants'

RETURN TO FLIGHT

APOLLO 7, SOYUZ 2, SOYUZ 3 AND ZOND 6

1968

The year 1968 saw a resumption of manned space flight by both the Soviet Union and the United States, the first time that either country had launched a human into space since the misfortunes of the previous year. Manoeuvres essential for a manned lunar landing – such as docking in space – preoccupied both nations, despite Soviet denials that their immediate goal was a mission to the Moon. However, once the bold plan by the Americans to attempt a manned lunar orbit were known, an opportunity arose for the Soviet Union to steal ahead in the race.

More by luck than planning, the Soviet 'return to flight' Soyuz mission would take place in time for the 51st anniversary of the Great October Revolution. The plan was to carry out a manned repeat of the successful automated docking of a year before – in other words, for a cosmonaut in a manned Soyuz to link up with an automated unmanned Soyuz. The two ships would remain docked for a few hours before separating and carrying out independent missions. Such a conservative rendezvous and docking mission would hopefully lead the way for the long-delayed intership cosmonaut transfer attempt. The Soviet political leadership was anxious to resume space missions after the long gap, particularly because of NASA's forthcoming *Apollo 7* mission in October – the first manned US spaceflight since the disastrous *Apollo 1* fire of January 1967.

Apollo 7

The redesigned Apollo capsule was launched as *Apollo 7* on 11 October 1968 from launch pad 34 at Cape Kennedy. On board were Wally Schirra, Donn Eisele and Walter Cunningham – the *Apollo 1* backup crew. It was not only the United States' return to flight after the tragedy, but an important shakedown flight to test the cone-shaped Apollo Command Module for the first time in space, along with its associated Service

Apollo 7 blasts off at 11.03 a.m. on 11 October 1968. A tracking antenna is seen on the left.

Module. It was also the first manned flight of a *Saturn* booster, in this case the *Saturn 1B* variant. It was 68 metres (224 feet) in height, and humans had never ridden into space on a more powerful rocket. Schirra was now 45 years old and making his third spaceflight. Alongside him were two rookies. Knowing that it was almost certainly his last trip into space, Schirra was determined that it should be a perfect mission – and especially his mission. Unfortunately, shortly after lift-off he developed a cold.

> Schirra: *We launched on a Friday. I remember this very specifically. In orbit, our so-called Friday night, Donn Eisele was on watch and Cunningham and I were supposed to be sleeping. And I hear Donn saying: 'Wally won't like that.' I put on my mike and listened in. 'Oh, we're supposed to put on the television tomorrow morning.' I said: 'Well, we didn't have it in the schedule, gentlemen. That doesn't go on till Sunday morning.' I should have said: 'I don't want to interrupt Howdy Doody [a popular television programme],' but I wouldn't have gotten away with it. What I really was saying was: 'We have not checked this system out. It's in the flight plan to be checked at this point in time. We'll check it at that point in time.'*

> Cunningham: *Well, Apollo 7 became very important. If we had not had a success on Apollo 7, we really don't know what would've happened to the space programme. Another accident and the fainthearted in the country, as we have a tendency to be, would've been clamouring to stop it. There was some real bickering back and forth between Wally and the ground. I, frankly, have never felt like I had any kind of a problem with the ground, but Wally was still demonstrating that it was Wally's flight and Wally was in charge. He has maintained since, that he felt the responsibility. He's never said that what he did was anything except the responsible thing to do. I really think it's a case of, in some instances, Wally wanting to insist he was in charge when nobody cared who was in charge anyway.*

The successful mission lasted almost 11 days. They simulated many of the events that would be required for a mission to the Moon. At one stage their rocket propelled them into a 269-mile- (433-km-) high orbit. 'That was a real kick in the pants,' exclaimed Schirra. Re-entry went according to plan, although Schirra refused to don his helmet for the procedure. The Apollo equipment received a thumbs-up, even if the commander of the flight did not.

> ## 'I really think it's a case of, in some instances, Wally wanting to insist he was in charge when nobody cared who was in charge anyway.'
>
> WALTER CUNNINGHAM

Soyuz 3 Runs Out of Fuel

Just a few days after *Apollo 7* returned, target vehicle *Soyuz 2* passed over the Russian launch site. And at that moment the USSR's return to flight, *Soyuz 3*, lifted off with Colonel Georgi Beregovoy aboard. It was the first-ever piloted launch from site 31, the second launch complex at the Baikonur Cosmodrome. At 47 years old, Beregovoi was at that time the oldest person to go into space. Once in orbit, the Igla automated docking system brought *Soyuz 3* to within 200 metres (656 feet) of *Soyuz 2*, at which point Beregovoi took over manual control. But the two ships were not aligned perfectly and instead of stabilizing his ship along a direct axis to *Soyuz 2*, Beregovoi put his spacecraft into an incorrect orientation. This caused *Soyuz 2*'s radar system, sensing an error, to automatically turn the craft's nose away to prevent an incorrect docking. Beregovoi did not see the problem and performed a fly-around, and then tried to approach *Soyuz 2* for a second time, but the same thing happened. By this time he had almost exhausted all the propellant available for such manoeuvres, meaning that further docking attempts had to be called off. For three days, 22 hours and 50 minutes Beregovoi had circled the Earth 64 times; while his flight may not have been successful, at least it was not a disaster.

1968

11 October Redesigned *Apollo 7*, with three astronauts on board, orbits Earth, marking an American return to space fights

14 October Soviet academician Leonid Sedov denies Russia plans to send cosmonauts to the Moon

26 October Russians launch *Soyuz 3*, which attempts unsuccessfully to dock in space with an orbiting *Soyuz 2* craft

10 November Soviets launch *Zond 6*, which successfully circles the Moon but crashes on landing back on Earth

Competing for the Next Mission

Because of delays to the next flight-ready L1 vehicle, the Soviets had to forego the October lunar launch window, thus shifting any possible launch into November. Soviet space planners were aware of the rumours of an Apollo lunar-orbital mission by the end of the year so they resorted to their usual public tactic – obfuscation – giving contradictory positions. On 14 October academician Sedov, who was representing the Soviet Union at the 19th Congress of the International Astronautical Federation in New York, stated:

> *The question of sending astronauts to the Moon at this time is not an item on our agenda.*
> *The exploration of the Moon is possible, but is not a priority.*

It was a lie.

The success of the *Apollo 7* mission crystallized an audacious idea that had already been discussed at NASA. It was in early August that George Low, the deputy director of NASA's Manned Spacecraft Center in Houston, ordered his staff to work on a plan to eliminate the so-called 'E' Apollo mission in favour of the much more ambitious 'C-prime' flight – in which an Apollo Command and Service Module launched on a *Saturn V* would go directly into lunar orbit. It was a risky decision, since it would be only the third launch of the *Saturn V* booster and the first time humans had flown on it, not to mention the obvious fact that the risks of going into lunar orbit were far greater than going into orbit around the Earth. But the advantages were many in terms of technical and scientific knowledge, as well as providing a demonstration of what the United States could achieve. A few weeks later NASA HQ gave its approval for the 'C-prime' mission, provided that *Apollo 7* was successful. Furthermore, *Zond 5* had already gone around the Moon, and as far as NASA knew a Soviet manned circumlunar flight could take place any time soon. The Soviets had a manned lunar launch window in December 1968. Would they be able to upstage *Apollo 8*?

By early November the Soviets were still planning two more automated L1 lunar missions, one in mid-November and one in early December, to be followed by a manned circumlunar flight in January. But once they heard about the planned *Apollo 8* lunar orbit, they realized they had an advantage if they could only use it. The *Apollo 8* launch window opened on 21 December, but because of different lunar trajectories undertaken from the two launch sites the circumlunar launch window for a Soviet launch from central Asia would occur earlier, around 8–10 December. However, despite much press speculation in the West, and an increase in tension approaching 8 December, the USSR was just not in a position to take advantage of the opportunity.

Circling the Moon, Crashing on Earth – *Zond 6*

An automated L1 launch did take place on 11 November, sending the spacecraft designated *Zond 6* towards the Moon. As soon as it was on its way controllers discovered that an antenna boom had not deployed. Despite this the mission went very well, with *Zond 6* flying around the far side of the Moon two days later

The lunar surface, with the Earth in the background,
photographed by the Soviet *Zond 8* spacecraft on 24 October 1970;
this was the last of the Zond unmanned circumlunar missions.

at a closest distance of 1500 miles (2420 km). After it had circled the Moon, controllers had to refine the spacecraft's trajectory for it to perform a guided re-entry into Earth's atmosphere and land on Soviet territory. The first correction was successfully accomplished, and it looked as if everything was on track until controllers detected a disastrous problem: the air pressure within the descent apparatus had dropped, indicating a compromise of the spacecraft's structural integrity. Despite the partial depressurization, later found to be the result of a faulty rubber gasket, the critical systems on the ship remained operational, and the controllers were able to carry out the third and final mid-course correction, just eight and a half hours prior to re-entry at a distance of 75,000 miles (120,000 km) from Earth.

On the morning of 17 November *Zond 6* separated into its two component modules prior to re-entry. Passing through its 5700-mile- (9000-km)-long re-entry corridor, it skipped out of the atmosphere, having reduced velocity down to 4.7 miles (7.6 km) per second, and began a second re-entry that further lowered velocity to only 200 metres (656 feet) per second. The complex re-entry was a remarkable demonstration of the precision of the L1 re-entry profile designed to reduce G forces. However, during part of the descent, pressure in the descent apparatus reduced further, killing any biological specimens on board. No doubt, a crew within the ship would have perished as well. Then the parachute system failed and it plummeted to the ground and smashed into pieces. Remarkably, the impact occurred only 10 miles (16 km) from the *Proton* launch pad at the Baikonur Cosmodrome, where *Zond 6* had lifted off just six days and 19 hours previously. The crushed descent apparatus clearly carried a lot of valuable materials. Among the items recovered intact from the wreckage was the exposed film from the camera, which provided beautiful pictures of both Earth and the Moon.

Because of the crash, Mishin postponed any plans for a piloted L1 mission in the near future; the dreams of Soviet engineers and scientists of circling the Moon prior to the United States were over. As the historic *Apollo 8* launch grew closer, Soviet spokespersons began to neutralize what was undoubtedly a public relations disaster. Veteran cosmonaut Titov, on a trip to Bulgaria, told journalists the day before the *Apollo 8* launch:

> It is not important to mankind who will reach the Moon first and when he will reach it – in 1969 or 1970.

But it did matter. It meant everything.

'Apollo 8, *you're go for TLI*'

LEAVING THE CRADLE

APOLLO 8
1968

Exciting, courageous and technically skilful though the events in space travel had hitherto been, if the United States could achieve the next phase of its Apollo programme, then it would eclipse all of these in one fell swoop. It planned to send astronauts on a voyage to another celestial body, the Moon – a journey that would necessitate humans leaving Earth's orbit for the first time in their history.

Apollo 8's three-man crew were mission commander Frank Borman, Command Module pilot James Lovell, and Lunar Module pilot William Anders. The mission also involved the first manned launch of a *Saturn V* rocket, and was the second manned mission of the Apollo programme.

Plan For the Mission

Originally planned as a low-earth orbit Lunar Module/Command Module test, the mission profile was changed to the more ambitious lunar orbital flight in August 1968. The overall objectives of the mission were to demonstrate Command and Service Module performance both between the Earth and Moon and in a lunar-orbit environment, to evaluate crew performance in a lunar-orbit mission and to return high-resolution photography of proposed Apollo landing areas and other locations of scientific interest.

> Frank Borman: *It's hard for us to fathom now, but the thing that's interesting about that mission was that, I don't know, maybe half a dozen of us sat in Chris Kraft's office one afternoon and we went over the flight plan, to try to understand what would we do on the whole flight. And I've always thought, again, it was an example of NASA's leadership with Kraft and their management style that we were able to hammer out, in one afternoon, the basic tenets of the mission.*

Jim Lovell had come into the mission to replace an injured Michael Collins, who had suffered severe back problems during training.

> Jim Lovell: *We were going to go out to 4000 miles so that we could test the Lunar Module, the Command Module, and then come back at a high rate of speed so that, you know, we could test*

The crew of *Apollo 8* (from left to right):
James Lovell, William Anders and Frank Borman.

the heat shield and things like that. I recall this very vividly. The three of us were out at Downey at North American testing our spacecraft; and Frank got a call to go back to Houston. So Bill Anders and I still stayed out there. We were working out there. And Frank came back again, back to Downey, and said: 'Things have changed.' And we said: 'Viz. a what?' He said: 'If everything goes all right with Apollo 7, we'll – Apollo 8 will go to the Moon.' I was elated! I thought: Man, this is great! I mean, I had already spent two weeks in space in Gemini 7 with Frank Borman. I didn't want to spend another 11 days, or something like that, you know, going around the Earth again. I said: 'This is fantastic!'

Gene Kranz: *Most of the people give the credit for* Apollo 8 *to a decision in August where George Low said: 'Hey, you know, I think, in order to keep this programme on track – we've got problems in the Lunar Module; it's behind schedule, it's overweight, there are software problems there – I think that we've got to go to the Moon.'*

The World Watches

When *Apollo 8* lifted off, the eyes of the world were upon the three astronauts. Kamanin wrote in his diary:

The flight of Apollo 8 *to the Moon is an event of worldwide and historic proportions. This is a time for festivities for everyone in the world. But for us, the holiday is darkened with the realization of lost opportunities and with sadness that today the men flying to the Moon are not named Valeri Bykovsky, Pavel Popovich or Aleksei Leonov, but rather Frank Borman, James Lovell and William Anders.*

The KGB (the Russian secret police) tried to stop *Apollo 8* from being launched by sending a letter to Cape Kennedy saying that the *Saturn V* had been sabotaged. US security officials saw through the ploy.

Borman: *I didn't want – really want – the mission to get fouled up because we really weren't certain that the Russians weren't breathing down our backs. So I wanted to go on time.*

Escaping the Bonds of Earth

Once in Earth orbit, the crew of the *Apollo 8* Command and Service Module had a decision to make – to fire the main engine that would take the spacecraft away from the Earth, making them the first humans to leave their home world and venture towards the Moon. Michael Collins, later to be a member of the historic *Apollo 11* crew, acutely observed that when *Apollo 8* left Earth's orbit for the first time, that event might in the long term be considered more important even than the first Moon landing.

Collins: *I think* Apollo 8 *was about leaving and* Apollo 11 *was about arriving, leaving Earth and arriving at the Moon. As you look back 100 years from now, which is more important, the idea that people left their home planet or the idea that people arrived at their nearby satellite? I'm not sure, but I think probably you would say* Apollo 8 *was of more significance than* Apollo 11, *even though today we regard* Apollo 11 *as being the zenith of the Apollo programme but ... historians may say* Apollo 8 *is more significant; it's more significant to leave than it is to arrive.*

But there was a problem. What do you say when you leave the Earth for the first time – so called trans-lunar injection (TLI)?

Collins: *I can remember at the time thinking: Jeez, there's got to be a better way of saying this, but we had our technical jargon, and so I said: 'Apollo 8, you're go for TLI.' If, again, 100 years from now you say you've got a situation where a guy with a radio transmitter in his hand is going to tell the first three human beings they can leave the gravitational field of Earth, what is he going to say? He's going to say something like – he's going to invoke Christopher Columbus or a primordial reptile coming up out of the swamps onto dry land for the first time, or he's going to go back through the sweep of history and say something very, very meaningful, and instead he says: 'What? Say what? You're go for TLI?' Jesus! I mean, there has to be a better way, don't you think, of saying that? Yet that was our technical jargon.*

Not that the Moon seemed to be getting any closer to Bill Anders:

We'd been going backwards and upside down, didn't really see the Earth or the Sun, and when we rolled around and came around and saw the first Earth rise, that certainly was, by far, the most impressive thing. To see this very delicate, colourful orb which to me looked like a Christmas tree ornament coming up over this very stark, ugly lunar landscape.

Christmas Message from Space

Apollo 8 entered lunar orbit on Christmas Eve, 24 December 1968. That evening, the three astronauts made a live television broadcast from lunar orbit, in which they showed pictures of the Earth and Moon seen from *Apollo 8*.

The reading of the Book of Genesis by the crew of *Apollo 8* whilst orbiting the Moon that Christmas – humans farther away from home than any had been before in history – is an iconic moment.

Letters flooded into NASA from all countries congratulating the crew of *Apollo 8* on their achievement. The year had been a bad one for the United States: the raging war in Vietnam; race riots; the assassinations of Martin Luther King and Senator Bobby Kennedy. Someone wrote to *Apollo 8* saying: 'Thank you for saving 1968.'

A Contradictory Response

In the USSR academician Sedov, still referred to as the 'father of the *Sputnik*', told Italian journalists a day after the *Apollo 8* splashdown that the Soviets had not been competing in a race to orbit or land on the Moon. Referring to *Apollo 8*, he added:

There does not exist at present a similar project in our programme. In the near future we will not send a man around the Moon, we start from the principle that certain problems can be resolved with the use of automatic soundings. I believe that in the next ten years vehicles without men on board will be the first source of knowledge for the examination of celestial bodies less near to us. To this end we are perfecting our techniques.

The Central Committee and the USSR Council of Ministers issued a new decree on 8 January 1969, 'On the Work Plans for Research of the Moon, Venus, and Mars by Automatic Stations'. Soon they would state in public that the USSR never wanted to go to the Moon at all. But behind the scenes it was different.

'We're actually going to fly something like this?'

DANGEROUS AND UNPLEASANT MISSIONS

SOYUZ 4, SOYUZ 5 AND APOLLO 9
1969

Despite the bravery of the Soviet cosmonauts, in 1969 the Soyuz programme, which included space docking and the transfer of crew from one craft to another, was overshadowed by the efforts of the American Apollo flights – but not before some near disasters had narrowly been averted and concerns had been voiced about the quality of the equipment being used. The year also saw the tragic death of a space legend.

On 13 January 1969 Vladimir Shatalov boarded the next Soyuz to be launched, *Soyuz 4*. Nine minutes prior to lift-off, the countdown was abruptly halted due to a failure in a hydraulic system and the launch was postponed 24 hours. It was another freezing day when launch operations began again. This time, there were no problems. Lt Colonel Vladimir Shatalov, 41 years old at the time, lifted off at 10.32 a.m. Moscow Time. In space, he manually fired the main engine of *Soyuz 4* on the fifth orbit to change his orbital parameters to 129 by 147 miles (207 by 237 km), to await the vehicle with which he would rendezvous.

Soyuz 4 and Soyuz 5 Dock in Space

The next day, *Soyuz 5* lifted off with its three-cosmonaut crew of 34-year-old Lt Colonel Boris Volynov (commander), 34-year-old civilian Aleksei Yeliseyev (flight engineer) and 35-year-old Lt Colonel Yevgeni Khrunov (research engineer). As soon as *Soyuz 5* reached orbit both spacecraft began their approach towards each other. In contrast to the original plans for the mission, which envisioned a docking on the very first orbit of the passive ship, the manoeuvres were carried out at a more leisurely pace over the period of a day. As Shatalov closed in on *Soyuz 5*, there were the seemingly inevitable problems, including erroneous signals from the docking control and spurious contact lights, but soon the two spacecraft hard-docked to Volynov's exclamation of 'Welcome!' Straight away Khrunov and Yeliseyev began their preparations for their transfer spacewalk by entering the living compartment of the *Soyuz 5* spacecraft.

The *Apollo* 9 crew of James McDivitt, David Scott and Russell Schweickart.

Each of their spacesuits had a self-contained life-support system attached to one of their legs instead of their backs. *Soyuz 5* carried letters addressed to Shatalov from his family and from ministers, as well as newspaper articles about his launch.

Spacewalk from *Soyuz 5* to *Soyuz 4*

Volynov bid the two cosmonauts goodbye and retreated back into the *Soyuz 5* Descent Module, closing the hatch between the two modules to allow depressurization. Khrunov then opened up the outer hatch of the living compartment on *Soyuz 4*'s 35th orbit and cautiously poked his head out. He recalled later:

> *I was amazed by the marvellous, magnificent spectacle of two spacecraft linked together high above the Earth. I could make out every tiny detail on their surfaces. They glittered brilliantly as they reflected the sunlight. Right in front of my eyes was* Soyuz 4.

Yeliseyev followed Khrunov, crawling towards the docking unit of *Soyuz 5*. They moved over to the living compartment of *Soyuz 4*, opened its hatch and crawled in. A welcome note from Shatalov, who was in the ship's pressurized descent apparatus, was waiting for them. After pressurization of the living compartment had taken place, the hatches between the two modules were opened, and Shatalov embraced his comrades, treating them to a toast of blackcurrant juice instead of the customary vodka. The entire episode had lasted one hour.

> ## 'Right in front of my eyes was Soyuz 4.'
>
> BORIS VOLYNOV

Wasting no time, the two commanders, Shatalov and Volynov, prepared for undocking. Just four hours and 34 minutes after docking, the two spacecraft separated and went on their own way, *Soyuz 4* now with three cosmonauts and *Soyuz 5* with one. They had been joined for three orbits. *Soyuz 4* was the first to return from orbit but Volynov, now alone, faced what was perhaps the most dramatic and dangerous re-entry in the history of the Soviet space programme.

Cheating Death – Twice

During the early morning of 18 January, in preparation for his re-entry around midday, Volynov reported that all systems on board *Soyuz 4* were fine. At 10.20 a.m., he passed over Africa before firing the engine for the predetermined period to initiate re-entry. Six seconds after the termination of retrofire, Volynov heard the pyrocartridges ignite, triggering the separation of the three major modules of the spacecraft: the living compartment, the Descent Module and the Service Module. However, when he looked out of the window for confirmation he immediately saw that something was wrong. He could see the antennas attached to the solar arrays on the cylindrical Service Module, meaning that it had not separated from the Descent Module. While similar failures had occurred on early Vostok and Voskhod flights, it posed a much greater threat on Soyuz because of the relatively large size of the module. Volynov immediately reported his predicament in code to ground controllers. Most believed he was doomed.

As the Soyuz descent apparatus – still attached to the 3-ton Service Module – entered the Earth's atmosphere, it began to perform a series of somersaults, exposing unprotected parts of the craft to the severe heat of friction. Smoke began to appear within the capsule. Normally during a re-entry, hydrogen peroxide thrusters would fire to provide lift and alter the craft's trajectory in order to reduce thermal and gravitational stresses. But although the instrument panel indicated that the valves for the thrusters were open, Volynov noticed there had been no firings – all the propellant had been used up during the retrofire phase, when the computer had tried in vain to correct the space ship's incorrect orientation.

Soviet cosmonaut Vladimir Shatalov explains how *Soyuz 4* and *Soyuz 5* docked in Earth orbit on 16 January 1969.

Volynov was sure he was only a few minutes from death. He considered saying goodbye to his family but decided to hurriedly save all the recorded materials on the docking procedure by ripping the important pages from the logbook, rolling them up tightly and placing them into the middle of the book. Then he calmly began to speak into a tape recorder, describing all the details of his experience to assist in identifying the reasons for the failure. There were more terrifying moments. Once, there was a sharp clap – the propellant tanks of the Service Module had blown apart; it happened with such force that the crew hatch was forced inwards and upwards like the bottom of a tin can. Volynov then realized that the troublesome Service Module had disintegrated, but his problems were not over yet. The straps on the main parachute began to twist, preventing the parachute from unfurling properly. For the second time he was sure he was going to die, but then the braids of the parachute began to untwist and the capsule landed, its soft-landing engines firing to slow the final descent. Even then, the landing was still so hard that the teeth in Volynov's upper jaw were broken off at the roots. Fortunately, the specially moulded couch saved him from broken bones and more serious injuries.

The Race is Over

In spite of the near catastrophe at the end of the flight, the *Soyuz 4–5* mission was a landmark flight in the Soviet space programme. It was not only the first docking of two piloted spacecraft in space but also the first transfer of a crew in orbit from one spacecraft to another. While the mission was nearly two years late it liberated them from the shadow of Komorov and gave them a success – but compared with the US space programme it was still relatively unimpressive. NASA astronauts had accomplished the first docking in space in March 1966 in *Gemini 8*. But after the humiliation caused by *Apollo 8*, the Soviet leadership was willing to take anything remotely successful as a triumph. They made much of the fact that the two spacecraft had been, in their words, the world's first 'experimental orbital station'. In the press conference Khrunov let it slip out that:

> In the design of our spacesuits certain aspects of Leonov's suit were taken into consideration.
> Our experiences on this flight may well contribute to the designs of a Moon suit.

While 1969 was to be the most dramatic year in the history of spaceflight, it was also perhaps the year when the Soviets realized that the space race was over and the political leadership intellectually abandoned their space programme. Space achievements were used as a means to sell the virtues of socialism in the early 1960s, but now Soviet officials were almost embarrassed by them, for they seemed to pale into insignificance compared with the achievements of the Americans. Against this backdrop, senior Soviet space officials convened in January 1969 to discuss not only an adequate response to the US space programme, but also to talk in general about the larger direction of their entire manned space effort. Deputy Chief Designer Chertok observed that the Soviet space programme had fewer resources than the US programme and yet was spending its money with even less rationality. It was an accurate observation on the state of the poor management of the Soviet space programme in the 1960s.

'Go on, go up, take off.'

BORIS CHERTOK

N1 Rocket Crashes to the Ground

The launch date for the first *N1* was set for February, but there were still signs that all was far from well in the form of an alarming report from Baikonur Cosmodrome commander Major General Aleksandr Kurushin. He refused clearance for the launch, due to many problems in the ground equipment and in the rocket itself. He only gave way after considerable pressure from most of the members of the State Commission. So it was that the doomed rocket was wheeled to the launch pad and levered upright.

In an attempt to take some of the gloss off the Apollo programme, the Soviets had built an unmanned Moon Rover. It was launched successfully on 19 February, but just 51 seconds after launch the payload abruptly fell apart, and the booster exploded. Two days later, almost four years late, the most powerful rocket ever built fired its 30 first-stage engines, producing 4590 tons of thrust. In just over ten seconds the *N1* left the pad. According to Boris Chertok:

> All the surrounding area shakes, there is a storm of fire, and a person would have to be insensitive and immoral to be able to remain calm at such moments. You really want to help the rocket: 'Go on, go up, take off.'

All seemed well until 70 seconds into the launch, when the control system abruptly shut down all the engines of the first stage, well before planned cut-off. Its momentum continued to carry the *N1* upwards to an altitude of 17 miles (27 km), before the rocket gradually came down to crash about 31 miles (50 km) from the launch site. The emergency rescue system saved the capsule, however, which landed some 20 miles (32 km) from the pad area. Another failure.

Apollo 9 – Dull but Dangerous

In some accounts, *Apollo 9* is considered as being a rather mundane flight in comparison with other Apollo missions, yet there are those who believe it may have been one of the more dangerous missions ever flown. It was certainly one of the most important to the programme. It was the first flight of the Lunar Module – the spacecraft that would actually land on the Moon – and the first complete test of the entire Apollo system.

Its three-person crew were Mission Commander Jim McDivitt, 39, Command Module Pilot Dave Scott, 36, and Lunar Module Pilot Rusty Schweickart 33. The flight was launched on 3 March 1969.

> McDivitt: *The Lunar Module was the key to the whole programme. And trying to get it light enough to fly was a real challenge. We got to the point where we were filing little blousons off of castings and things like that to get the weight down. The main thing was that we got a chance to fly the Lunar Module to see if it really worked. We had a few minor glitches on the descent engine, as I recall, when we first started it up, but it worked fine. The rendezvous worked okay. The computers worked. The radar worked. I mean, we did a damn good job of engineering it, because we really didn't have very many big problems with the spacecraft. It all went together well.*

Would *Apollo 9* Make It?

It was an encouraging mission. There remained nine months before President Kennedy's deadline expired, but there were times when the crew of *Apollo 9* thought that the equipment wouldn't make it.

> McDivitt: *The first time Rusty and I went up to Grumman to do a storage review – a storage review is something you do before you really solidify where you're going to put everything, and it's so you can still make some changes. And we went over to this vehicle sitting there in the corner, and we had two different kinds of vehicles. We had non-flight vehicles that were heavy construction; we had the flight-weight vehicles. And we go over there, and we get in the spacecraft, and we crawl in. And I can remember the first thing we did is we knocked off the shield around the environmental control system, which was a thing about as thick as a piece of paper and made out of plastic. And so, we get in there and we start checking the stowage. We weren't checking the spacecraft. We were just checking to see if everything fitted. Every time we turned around,*

something else broke! And I'm pretty mild-mannered and I don't get excited when things aren't going right. But after we were doing this for about five or six hours, and everything we touched fell off the wall or broke or it did something! Finally I got on the radio and I said: 'Damn it, you guys! We've been here all day long. We're – and we've got this crappy training vehicle out here that, you know, we ought to get something that more resembles what the heck we're going to fly with in space instead of this junk that we've got here!' And then I shut up, and there's this long pause. And finally somebody comes on the intercom and says: 'Jim, that is the flight vehicle.' I looked at Rusty and he looked at me, and we said, 'Oh my God! We're actually going to fly something like this?'

During the mission we did a docked burn with the Lunar Module. We did a bunch of oscillating tests with the Command Module. We did an EVA. We checked all the alternative methods of doing star alignments. We had multiple burns on the descent stage. Throttled the engine up and down through regimes it probably was never throttled at when it landed in the Moon. And it worked out really great.

Sickness in Space

Russell 'Rusty' Schweickart joined NASA as one of 14 astronauts named in October 1963, the third group of astronauts selected. During the *Apollo 9* mission it was planned he should make an EVA to test the portable life-support backpack, which was subsequently to be used on the lunar surface explorations.

Schweickart: Although the Russians had had a fairly strong record that people would get sick in space, up until Apollo 8 we had actually not experienced it. Our spacecraft were smaller, we're much more restricted, and on Apollo 7 no one reported any problem. But on Apollo 8, Frank Borman had gotten sick, but for all kinds of reasons, which are Frank's, he wouldn't really come forward with it. He didn't do any tests afterward. So we didn't know a lot about it, but I was fairly cautious because I would get sick on the zero G airplane, our affectionately dubbed Vomit Comet, which I'm sure you've heard about before. After successive parabolas, it's a very sickening experience and a challenging one for motion disturbance. So I knew from that and from a few episodes of seasickness that I was susceptible to motion sickness.

Well, of course, we didn't know that, so then on the third day of the mission, when it's time for me to go into action, that's really the first time when I'm moving around, and the first thing I've got to do is get into the space suit early in the morning. Getting into the space suit is a real contortionist challenge. So I got into the space suit, in which you have to double over. I mean, it's very interesting. I used to love to do it, and I could get into the suit as well as anybody or better than almost all people because I was also doing some of the early suit work.

When I popped my head through the suit and stood up and started zipping up the suit, I was not feeling too well. This was early in the morning before getting ready to go into the Lunar Module. I sort of slowed down to try and take it easy, but once that process of malaise starts

This photograph of *Apollo 9* Command Module Pilot
Dave Scott standing in the open hatch was taken
by Rusty Schweickart during his EVA.

going, you know, it kind of has a natural dynamic. So suddenly I had to barf, and I'm grabbing
for a bag, barfed in the bag, and, I mean, that's not a good feeling. But, of course, you feel better
after you barf, like anytime you get motion sickness, you feel better after it, but you don't like to do
it. Of course, that was sort of a warning shot. I mean, you know, oooh, we got a problem here?

'I mean, you know, oooh, we got a problem here?'

RUSSELL 'RUSTY' SCHWEICKART

So then I go over into the Lunar Module, and that's also a challenge, because now you're going from an environment that you're used to in the Command Module where that's up, now you go over into the Lunar Module. Well, in the Lunar Module, you're used to that being up. But now they're 180 degrees. So you're used to being up, and when you go over there, it's down. So you're having to change axes and do all kinds of stuff. So when I went over there to activate the Lunar Module, I was moving very slowly and deliberately and using my eyes a lot, and trying to keep my head from moving, because I sure as heck didn't want to get sick again. That worked out. After I got things turned on, then McDivitt came over and we started working together. We're slowly moving toward the afternoon activities, and at one point we're both busy.

All of a sudden I had to barf again, so then twice, for the second time that day, I'm grabbing for a bag and I barf. Again, after it's over you feel better, but now I've barfed twice and, of course, we're all very aware of that. Even though I'm feeling better immediately after I barfed, I'm still not feeling great. So we get everything done. It didn't delay us at all, but we got everything done, but we got back in the spacecraft and now the question is, you know, the next day I've got to go EVA, or scheduled to go EVA. Well, barfing in space is no fun. Jim decides: 'Well, we'll cancel the actual EVA tomorrow. We'll go right up to the point of depressurizing the hatch.' I can still have the helmet on, do all the checkout on the portable life-support system, the whole thing. But when it comes time to actually depressurize the Lunar Module, we'll simulate depressurizing, assume that you've been outside, you've come back in, and we've just repressurized. Then we'll pick up all of the checklists and everything from there. So we'll get all the tests and checkout and all the procedures and make sure every — we'll get as much done as we possibly can, but we're not actually going to do the EVA.'

Well, then it's time to go to sleep and get ready for the next day, store everything. And again I'm still not feeling very well. Of course, now I've just been the cause of not doing the EVA, which means that the portable life-support system really isn't checked out the way it was supposed to be checked out, so it's not really ready for the lunar surface missions, and are we going to run into some problems? This is already March of 1969. That end of the decade, I mean, is coming right up. Am I going to get so sick that we have to — am I going to remain sick, or are we going to have to actually abort the mission and the whole rest of the mission, in fact? Are we not going to be able to do the rendezvous? Is this basically a wasted mission because Schweickart's barfing? I mean, that's all going through my mind as I'm trying to go to sleep that night.

Well, the next morning I felt a lot better, and we were going over in the Lunar Module and getting ready for doing what we decided to do, started getting everything ready and moving around and checking this out and that out and the other thing. And I'm feeling considerably better. So somewhere maybe an hour before we were scheduled for the EVA, at that point Jim looks at me and I'm looking at Jim, and we're obviously thinking the same thing. He says: 'You know, you're looking a lot better today. How are you feeling?' I said: 'I'm feeling a lot better.' He said: 'It looks like it.' So we kind of looked at each other and said: 'Well, let's just keep going and we'll see what happens.'

So we go through probably another 45 minutes, maybe half an hour before the scheduled EVA, 15, 20 minutes before it, and nothing's changing because we're doing everything as if I'm going out anyway. So somewhere down there, 15 to 30 minutes or something like that in the records, we look at each other again and Jim says: 'How are you feeling?' I said: 'I'm feeling real good.' He says: 'You think you're okay?' I said: 'I think it's fine.' And he looked at me. We knew each other well enough, and he said: 'Okay, let's do it.' 'Right.' Jim calls the ground: 'We're going out on EVA.' Surprised them. That was just in 12 hours going from as low as I've ever been to about as high as I've ever been.'

Death of a Hero

On 27 March Yuri Gagarin, now Director of Training at Star City but who had not completely given up hope of flying in space once more, undertook a routine Mig-15 training flight out of Chkalovsky Air Base near Star City, along with flight instructor Vladimir Seregin. In circumstances that have never been fully explained, they crashed and were both killed. For some reason they had gone into a spin, possibly due to a near miss with another jet, and because an out-of-date weather report gave them a falsely high reading of their altitude, they were unable to gain control of the aircraft in time and it crashed.

'We is down among 'em Charlie'

REHEARSAL FOR A MOON LANDING

APOLLO 10
1969

Apollo 10 **was the fourth manned mission in the Apollo programme and a dress rehearsal for a landing on the Moon. Initially, it was thought that** Apollo 10 **would be the first mission to attempt the landing, but difficulties in the production of the Lunar Module meant that it would be left to** Apollo 11 **to undertake the task.**

On 18 May 1969 the Apollo 10 crew of Tom Stafford, John Young and Gene Cernan set off on their mission, the primary aim of which was to test a Lunar Module that would be capable of landing on the Moon. They were also to survey the Apollo 11 landing site in the Sea of Tranquillity. Apollo 10 added another first by broadcasting live colour television from space.

> Stafford: We were told that because of the trajectory we would fly, we would not see the Moon until we got there. And that's kind of weird, looking around. You see the Earth go by every 20 minutes, see the Sun go by. Where in the hell is the Moon? And the way the trajectory was and everything, the Moon was eclipsed. Finally, we got – well, a few hours out from the Moon, you could maybe see one little rim of it, hardly a rim. But most of the way to the Moon, we never saw it. Now, we saw it all the way back. Then the Earth started to be eclipsed, and just before we came back to see the Earth, all you could see was a little thin blue line of the Earth. So it was kind of unique. and right within a second – BOOM – the Earth goes down. The Earth disappears. There's this big black void. Down below the Earth, when it goes night time, you always see lights and cities and gas fires. There's just all kinds of lights around, and lightning all over. Just a big black void. So we left the Earth. It disappeared. It was quiet. Got turned around, and suddenly – couldn't see anything, and suddenly, about 60 seconds, we were all set, just counting down. Right below us, here comes the Moon, right out in daylight. So it was a real funny feeling there. It really looked weird. And to me, the colour of the Moon in early morning and late at night always looked a little reddish tinge on the top of the mountains. Some people say it's always white and black. I thought it was reddish, with maybe some charcoal greys and tans. So we went through the procedures, got a little bit of rest, and then got squared away.

The prime crew of *Apollo 10* (from left to right):
Gene Cernan, John Young and Thomas Stafford.

We got all squared away and started our manoeuvre to go down to about nine miles above the mountains and do two low passes, check out the landing radar, because if the landing radar doesn't work to update your state vector, you couldn't land. And it turned out the radar locked on to the lunar surface way in excess of spec, which was good. So as to what we did, we undocked and went way up high above him and came down low to get phasing behind him in case we had to abort to come up. So we did that, went down. What always amazed me was the size of the boulders. They were awesome, these big ones, you know, huge things. Some of them are pure white with black striations up on the side of these gigantic craters. I said, oh, they'd have to be as big as a two- or three-story building. It's hard to judge distance. Here on the Earth, even from space, you can still see some roads and you can see cities. You can kind of judge some distance. No roads up there.

They fired Snoopy's (the Lunar Module) rocket to drop down to within 15,240 metres (50,000 ft) of the Sea of Tranquillity. As they made their first pass over the southwestern corner of the Sea of Tranquillity, an excited Cernan called out: 'I'm telling you, we are low. We're close baby! We is down among 'em, Charlie.' Capcom Charlie Duke responded: 'I hear you weaving your way up the freeway.'

Stafford: We were all set to stage off, and I noticed the thrusters started to fire. I looked down and I could see I had a yaw rate, but I could tell by the eight ball I wasn't yawing. So I talked to Cernan, and started firing again. We were all buttoned up, and I started troubleshooting, went to the AGS [phonetic] position and all that, but the first thing you know – BOOM – the whole damned spacecraft started to tumble and tried to rotate like that. And real fast, I just reached over and just blew off the descent stage, because all the thrusters were on the ascent stage, get better torque-to-inertia ratio, because we're heading over towards gimble lock on the main platform.

Delusions of the Soviet Cosmonauts

The Soviets were unable to perform such space missions. None of their lunar spacecraft was near flightworthy, and the Soyuz spacecraft was more dangerous that it should have been. But despite dampening enthusiasm, a group of cosmonauts continued to prepare for lunar landings at the Yuri Gagarin Cosmonaut Training Centre and the Gromov Flight-research Institute. It was the cosmonauts, rather than the politicians or the engineers, that kept the Soviet lunar dreams alive. Aleksei Leonov said in the spring of 1969:

The Soviet Union is also making preparations for a manned flight to the Moon like the Apollo programme of the United States. The Soviet Union will be able to send men to the Moon this year or in 1970.

But Kamanin wrote in his diary during the *Apollo 10* mission of the 'unrestrained lying' by Soviet officials about their intentions with respect to the Moon. He added bitterly:

We have come to the end to drink the bitter chalice of our failure and be witnesses to the distinguished triumph of the USA in the conquest of the Moon.

Soviets Aim for the Moon

Throughout the spring and early summer of 1969, as the world waited for the first manned landing on the Moon, there was speculation that the USSR was planning something spectacular to upstage it. In reality the Soviets could do little. Even so, Wernher von Braun said that it was still possible for the USSR to reach the Moon before the United States if *Apollo 11* was delayed, and he strongly believed that the Soviets would undertake piloted lunar flight in the latter part of the year using a giant rocket. But von Braun was 'out of the loop', and in fact the US intelligence community knew the Soviets had no chance. A top secret CIA 'National Intelligence Estimate' of June stated that it would probably be 1972, and late 1970 at the earliest, before they could attempt a manned lunar flight. Von Braun also warned that a Soviet robotic spacecraft could bring back lunar soil before *Apollo 11* came back with its samples. In fact, the Soviet unmanned lunar sampler mission did indeed have two launch windows to reach the Moon in June and July of that year.

On 4 June a spacecraft was launched from Baikonur in a desperate move to obtain some lunar soil without putting human life at risk. If the gamble paid off, it would at least be some sort of achievement.

'I'm telling you, we are low. We're close baby!'

GENE CERNAN

But as the third stage of the *Proton* rocket burnt out and the Block-D was due to take over, a control system failure prevented it from firing. The mission was lost. Instead of going to the Moon, the lunar sampler ended up in the Pacific. The Soviets had four remaining lunar scoop spacecraft left and only one chance to beat *Apollo 11*. Things looked bleak; the *Proton* rocket had failed on all of its last five missions.

The mighty *N1*, which the Soviets hoped would eventually put a cosmonaut on the Moon, was moved to the launch pad with lift-off set for 3 July – under two weeks before *Apollo 11* attempted the first Moon landing. Before midnight the *N1*'s 30 first-stage rockets burst into life. Lieutenant Menshikov recalls:

> We were all looking in the direction of the launch, where the 100-metre pyramid of the rocket was being readied to be hurled into space. Ignition. The flash of flame from the engines, and the rocket slowly rose on a column of flame. And suddenly, at the place where it had just been, a bright fireball. Not one of us understood anything at first. There was a terrible purple-black mushroom cloud, so familiar from the pictures from the textbook on weapons of mass destruction. The steppe began to rock and the air began to shake and all of the soldiers and officers froze.

There was a deathly silence as the onlookers awaited the arrival of the blast wave. Menshikov recalls:

> Something quite improbable was being created all around – the steppe was trembling, thundering, rumbling, whistling, gnashing, together in some terrible, seemingly unending cacophony. The trench proved to be so shallow and unreliable that one wanted to burrow into the sand so as not to hear this nightmare. The thick wave from the explosion passed over us, sweeping away and levelling everything. Behind it came hot metal raining down from above.

Pieces of the rocket were hurled up to 6 miles (10 km) away, and large windows were shattered 25 miles (40 km) away. The 400-kilogram (880-lb) spherical tank landed on the roof of a building 4.4 miles (7 km) from the launch pad. Kamanin wrote in his diary:

> Yesterday the second attempt to launch the N1 rocket into space was undertaken. I was convinced that the rocket would not fly, but somewhere in the depth of my soul there glimmered some hope for success. We are desperate for a success, especially now when the Americans intend in a few days to land people on the Moon.

There was only one card left to play – the final chance to launch an unmanned sample-retrieving return mission. After five straight failures of the *Proton* rocket it finally performed well, lifting their last hope off the pad three days before the scheduled launch of *Apollo 11*. They called it *Luna 15*. The Soviet media said its mission was merely to study circumlunar space.

'Houston. Tranquility Base here. The Eagle has landed'

FIRST LANDING ON THE MOON

APOLLO 11
1969

Between 16 and 24 July 1969, the world held its breath and watched spellbound as one of humankind's most significant and audacious endeavours took place – the first attempt to land a human being on another celestial body. Despite the setbacks and concerns of earlier missions, *Apollo 11* was a triumph for all involved. Yet even during this momentous event, the Soviet Union put in place a plan to try to upstage the American effort.

Neil Armstrong was told that his mission, *Apollo 11*, would be the first to attempt a lunar landing. He recalls:

> *During the flight of* Apollo 8 *I had three or four meetings with Deke Slayton about, first, would I take the third one down to the surface and then we had a lot of talks about who might be available and be right to be on that crew, that sort of thing.*

The crew of *Apollo 11* – Neil Armstrong, Edwin 'Buzz' Aldrin and Michael Collins – were introduced to the press on 9 January 1969, and immediately the assembled reporters got down to the big question: 'Which of you gentlemen will be the first man to step out onto the lunar surface?' Over the years Aldrin has said that he would have preferred to have flown on a later mission. Writing in *Return to Earth* he said:

> *I would have preferred to go on a later flight. Not only would there be considerably less public attention, but the flight would have been more complicated, more adventurous, and a far greater test of my abilities than the first landing.*

First Man on the Moon?

It is clear that for the first few months of 1969 Aldrin believed he would be first out of the Lunar Module. He said he had never given it much thought and that he had naturally presumed that he would be first. After all there were precedents, beginning with Ed White's spacewalk, when the commander of the flight stayed in the spacecraft while his partner carried out the excursions. Aldrin was perhaps right to believe

'Buzz' Aldrin on the Moon in 1969. Behind him is the Lunar Module Eagle. To his right, the Solar Wind Composition experiment has been deployed.

1969

9 January The US astronauts who will attempt a Moon landing – Neil Armstrong, 'Buzz' Aldrin and Michael Collins – are introduced to the press

14 April It is announced that Neil Armstrong will be first astronaut out of the Lunar Module after the Moon landing

16 July *Apollo 11* blasts off from Cape Kennedy, destined for the Moon

20 July Lunar lander Eagle sets down on the Moon's Sea of Tranquillity. A few hours later, Armstrong, followed by Aldrin, leave their spacecraft to walk on the surface

21 July *Luna 15*, an unmanned Soviet craft sent to the Moon in an attempt to upstage the American endeavour, crashes into a lunar mountain

24 July Crew of *Apollo 11* return and splash down safely in the ocean

it; NASA's Associate Administrator for Manned Space Flight told several people, including several members of the press, that he would be the first on the Moon.

Word began to filter out, however, that it would be Armstrong. Armstrong was a civilian. Buzz was angry, considering it an insult to the service. He was technically still a member of the air force, although he had not served for ten years except to maintain his flying hours. So Aldrin approached Armstrong about the issue. Aldrin wrote later (although he claimed it was done by his co-author):

He equivocated a minute or so, then with a coolness I had not known he possessed he said that the decision was quite historical and he didn't want to rule out the possibility of going first.

Armstrong says he cannot remember that conversation. Aldrin talked to his colleagues, but for some of them that was seen as lobbying behind the scenes to be the first. Gene Cernan, quoted in Armstrong's biography, says:

He came flapping into my office at the Manned Spaceflight Center one day like an angry stork, laden with charts and graphs and statistics, arguing what he considered to be obvious – that he, the Lunar Module pilot, and not Neil Armstrong, should be the first down the ladder on Apollo 11. Since I shared an office with Neil Armstrong, who was away training that day, I found Aldrin's arguments both offensive and ridiculous. Ever since learning that Apollo 11 would attempt the first Moon landing, Buzz had pursued this peculiar effort to sneak his way into history, and was met at every turn by angry stares and muttered insults from his fellow astronauts. How Neil put up with such nonsense for so long before ordering Buzz to stop making such a fool of himself is beyond me.

Later Aldrin said he had given the wrong impression and that he really did not want to be first. It was up to Deke Slayton to put a stop to the talk. Slayton said that as Armstrong was a member of the second intake of astronauts, the group before Aldrin joined, he should have priority. Aldrin later said he was fine with that but felt uncomfortable that nobody else knew.

Announcing the Decision

On 14 April the speculation came to an end. At a press conference George Low said: 'The plans call for Mr Armstrong to be the first man out after the Moon landing. A few minutes later Colonel Aldrin will follow.' Aldrin later said he believed that the physical layout of the Lunar Module dictated that Armstrong went out first. Aldrin, the Lunar Module pilot, was on the right. But that was not the case. The layout was not the reason. In his biography Armstrong said:

> *In my mind the important thing was that we got four aluminium legs safely down on the surface of the Moon while we were still inside the craft. But it could technically have been Buzz. Just move before you put the backpacks on.*

Later Chris Kraft explained NASA's thinking. He said that they knew damn well that the first guy on the Moon was going to be a Lindbergh. Neil was calm, quiet and had absolute confidence:

> *We knew he was the Lindbergh type. He had no ego. The most he ever said about walking on the Moon was that it might have been that he wanted to be the first test pilot to walk upon the Moon. If you would have said to him, you are going to be the most famous human being on Earth for the rest of your life, he would have answered that he didn't want to be the first man on the Moon. On the other hand, Aldrin desperately wanted the honour and wasn't quiet in letting it be known. Neil said nothing.*

The *Apollo 11* astronauts (from left to right): Neil Armstrong, Michael Collins and Edwin 'Buzz' Aldrin.

Kraft said that nobody criticized Buzz but that they did not want him to be the man who would become legend: 'The hatch design didn't come into it. That was a rationalization, a solace for Buzz.'

Towards the Lunar Surface

The Lunar Module – Eagle – with Armstrong and Aldrin on board, separated from the Command Module – Columbia – with Michael Collins on board, on the far side of the Moon 100 hours and 12 minutes into the *Apollo 11* mission. Armstrong said: 'The Eagle has wings.' Shortly afterwards they fired the Eagle's descent engine for 30 seconds to put them on the path to the surface.

> Kranz: *The spacecraft is now behind the Moon, and the control team, the adrenaline, I mean, just really was – no matter how you tried to hide it, the fact is that you were really starting to*

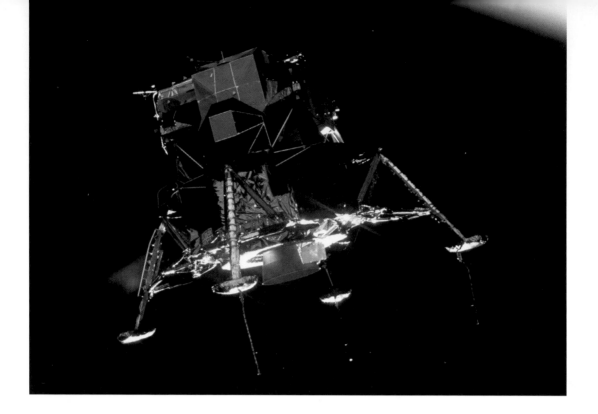

The Lunar Module Eagle on its
historic journey to the Moon's surface.

*pump. It seems that every controller has a common set of characteristics, is they've got to go to the
bathroom. I mean, it's just to the point where you just need this break. That's all there is to it. It's
literally a rush to get to the bathroom. You're standing in line, and for a change, there isn't the
normal banter, no jokes, etc. I mean, the level of preoccupation in these people – and these are
kids. The average age of my team was 26 years old. Basically I'm 36; I'm ten years older. I'm the
oldest guy on this entire team.*

*This day, is either going to land, abort, or crash. Those are the only three alternatives. So it's
really starting to sink in, and I have this feeling I've got to talk to my people. The neat thing about
the Mission Control is we have a very private voice loop that is never recorded and never goes
anywhere. It's what we call AFD (Assistant Flight Director) Conference Loop. It was put in there
for very specific purposes, because we know that any of the common voice loops can be piped into
any of the offices at Johnson. They can be piped into the media, they can be piped into the viewing
room, and what we want is an incredibly private loop where we can talk to the controllers when
we need to, but in particular it was set up for debriefing, because debriefings are brutal.*

*So I called the controllers, told my team: 'Okay, all flight controllers, listen up and go over to
AFD Conference.' And all of a sudden, the people in the viewing room are used to hearing all
these people talking, and all of a sudden there's nobody talking any more. But I had to tell these
kids how proud I was of the work that they had done, that from this day, from the time that they
were born, they were destined to be here and they're destined to do this job, and it's the best team
that has ever been assembled, and today, without a doubt, we are going to write the history books*

and we're going to be the team that takes an American to the Moon, and that whatever happens on this day, whatever decisions they make, whatever decisions as a team we make, I will always be standing with them, no one's ever going to second-guess us. So that's it.

We can't communicate to them; they can't communicate to us. The telemetry is very broken. We have to call Mike Collins in the Command Module to relay data down into the Lunar Module, and immediately this mission role has come into mind because it's decision time, go/no go time. It just continues, broken, through about the first five minutes after we've acquired the data, but we get enough data so the controllers can make their calls, their decisions. Are we good? Are we properly configured? Are we basically at the point in the procedures where we should be?

We move closer now to what we call the 'powered descent go/no go.' This is where it's now time to say are we going down to the lunar surface or not. Now, I have one wave-off opportunity, and just only one, and if I wave off on this powered descent, then I have one shot in the next revolution and then the lunar mission's all over. So you don't squander your go/no gos when you've only got one more shot at it.

We lose all data again. So I delay the go/no go with the team for roughly about 40 seconds, had to get data back briefly, and I make the decision to press on; we're going to go on this one here. So I have my controllers make their go/no gos on the last valid data set that they had. I know it's stale, but the fact is that it's not time to wave off. So, each of the controllers goes through and assesses his systems right on down the line.

We get a go except for one where we get a qualified go, and that's Steve Bales down at the guidance officer console, because he comes on the loop, and he says: 'Flight, we're out on our radial velocity, we're halfway to our abort limits. I don't know what's caused it, but I'm going to keep watching it'. So all of a sudden, boom! We've sure got my attention when you say you're halfway to your abort limits. We didn't know this until after the mission, but the crew had not fully depressed the tunnel between the two spacecrafts. They should have gone down to a vacuum in there, and they weren't. So when they blew the bolts, when they released the latches between the spacecraft, there was a little residual air in there, sort of like popping a cork on a bottle. It gave us velocity separating these two spacecraft. So now we're moving a little bit faster by the order of fractions of feet per second than we should have at this time. So we don't know it, but this is what's causing the problem. It's now a problem.

In the meantime, we've had an electrical problem show up on board the spacecraft, and we've determined that this is a bad meter that we've got for the AC instrumentation. AC, alternating current, is very important on board the spacecraft, because it powers our gyro's landing radar right on down the line. We're now going to be looking at this from the standpoint of the ground so that Buzz won't have to look after it.

All through this time, my mind is really running. Is this enough data to keep going, going, going, going? Because I know what I'm going to do in this role. I'm going to be second-guessed, but that isn't bothering me. We now get to the point where it's time to start engines. We've got *telemetry back again. As soon as the engine starts, we lose it again. This is an incredibly important time to have our telemetry because as soon as we get acceleration, we settle our propellants in the tanks, and now we can measure them, but the problem is, we've missed this point. So now we have to go with what we think are the quantities loaded pre-launch. So we're now back to nominals. So we're in the process of continuing down.*

Computer Problems, Communication Problems

The capcom for the landing was Charles Duke, requested by Armstrong. Duke recalls: 'The communication dropouts were a nuisance more than a danger, but a computer problem was a showstopper.'

Armstrong: *You're always concerned when any kind of alarm comes on, but it wasn't a serious concern because there wasn't anything obviously wrong. The vehicle was flying well, it was going down the trajectory we expected, no abnormalities in anything that we saw, other than the computer said: 'There's a problem, and it's not my fault.' The people here on the ground were right on top of that, and of course, the computer continued in a contrary manner periodically all the way to the surface. But my own feeling was, as long as everything was going well and looked right, the engine was operating right, I had control, and we weren't getting into any unusual attitudes or things that looked like they were out of place, I would be in favour of continuing, no matter what the computer was complaining about.*

Kranz: *So now we're fighting – we've got this new landing area that we're going to be going into, we're fighting the communications, we've got the problem with the communications, and we've got the AC problem that we're now tracking for the crew, and now a new problem creeps into this thing, which is this series of programme alarms. There's two types of alarms. These are the exact ones that we blew in the training session on our final training day, twelve-oh-one. Twelve-oh-one is what we call a bail-out type of alarm. It's telling us the computer doesn't have enough time to do all of the jobs that it has to do, and it's now moving into a priority scheme where it's going to fire jets, it's going to do navigation, it's going to provide guidance, but it's basically telling us to do something because it's running out of time to accomplish all the functions it should.*

We tell them we're going the alarms. We tell them to accept radar, go on the alarms, you know, radar's good, getting close – you know, we're continuing to work our way down to the surface. Now, fortunately the communications have improved dramatically. Communications are no longer a concern of mine, but they were for about the first six or eight minutes of our descent. But now we're about four minutes off the surface. Communications are just a dream.

Hovering above the lunar surface, Armstrong looked for a landing site. He had taken over manual control at about 152 m (500 feet), and the first thing he did was to slow the rate of descent whilst maintaining his forward speed. There were huge blocks and an extensive boulder field below him. They could not land there, they had to press on. Ahead there appeared to be a more open area.

Cutting it Fine

Armstrong: *We could have tried to land there, and we might have gotten away with it. It was a fairly steep slope and it was covered with very big rocks, and it just wasn't a good place to go. You know, if I'd run out of fuel, why, I would have put down right there, but if I had any choice of a more promising spot, I was going to take it. There were some attractive areas far more level, far less occupied by boulders and things, a half mile ahead or so, so that's where I went.*

Duke: *When they pitched over to look at the lunar surface, they didn't recognise anything and they were going into this big boulder field and Neil was flying a trajectory that we'd never flown in the simulator. It was something we'd never seen. And, you know, we kept trying to figure out, 'What was this going? What's going on?' You know, he's just whizzing across the surface at about 400 feet, and all of a sudden he — the thing rears back and he slows it down and then comes down. And I'm sitting there, sweating out.*

Kranz: *Some person — and we've never been able to identify it in the voice loop — comes up and says: 'This is just like a simulation,' and everybody relaxes. Here you're fighting problems that are just unbelievable and you keep working your way to the surface, to the surface, to the surface. So we get down to the point — and we know it's tough down there, because the toe of the footprint is really a boulder field, so Armstrong has to pick out a landing site, and he's very close to the surface. Instead of moving slowly horizontal, he's moving very rapidly, and ten and 15 feet per second, I mean, we've never seen anybody flying it this way in training.*

Now Carlton calls out: 'Sixty seconds,' and we're still not close to the surface yet, and now I'm thinking: 'Okay, we've got this last altitude hack from the crew, which is about 150 feet, which now means we've got to average roughly about three feet per second rate of descent, and I see Armstrong's at zero. So I say: 'Boy, he's going to really have to let the bottom out of this pretty soon.'

Once Armstrong had chosen his spot it was necessary to lower the Lunar Module quite slowly. They got to within 15 m (50 feet) of the surface and inwardly Armstrong knew they had done it. Later Kranz said: 'I never dreamed we would still be flying this close to empty. When Duke called: "Thirty seconds," Neil wasn't worried about the fuel. They landed the simulators with 15 seconds of fuel left all the time.'

Armstrong: *There was a lot of concern about coming close to running out of fuel, and I was very cognisant of that. But I did know that if I could have my speed stabilized and attitude stabilized, I could fall from a fairly good height, perhaps maybe 40 feet or more in the low lunar gravity, the gear would absorb that much fall. So I was perhaps probably less concerned than a lot of people watching down on Earth.*

Then Aldrin said: 'Contact light.' These were the first words ever said on the lunar surface, spoken as a probe from one of the legs registered with the ground. A few seconds later Armstrong said: 'Shutdown.'

Jubilant ground controllers celebrate the successful conclusion of the *Apollo 11* landing mission.

Kranz: *Well, what happens, we have a three-foot-long probe stick underneath each of the landing pads. When one of those touches the lunar surface, it turns on a blue light in the cockpit, and when it turns on that blue light, that's lunar contact, their job is to shut the engine down, and they literally fall the last three feet to the surface of the Moon. So you hear the: 'Lunar contact,' and then you hear: 'ACA [Attitude Control Assembly] out of Detent [out of centre position].' They're in the process of shutting down the engine at the time that Carlton says: 'Fifteen seconds,' and then you hear Carlton come back almost immediately after that 15 seconds call and say: 'Engine shutdown,' and the crew is now continuing this process of going through the procedures, shutting down the engine.*

Duke: *Everybody erupted in Mission Control and then his famous lines about, 'Houston. Tranquillity Base here. The Eagle has landed.' And so we made it, you know, and it was really a great release. People cheering and all. I was so excited, I couldn't get out 'Tranquillity Base.' It came out sort of like; 'Twangquility.' And so it was: 'Roger, Houston. Twangquility Base here.' Let's see, what did I say? No, it was: 'Roger, Twangquility Base. We copy you down. We've got a bunch of guys about to turn blue. But we're breathing again.' And I believe that's true – was a true statement. It was spontaneous, but it was true. I mean, we were – I was holding my breath, you know, because we were close.'*

Kranz: *In the meantime we're just busier than hell doing our stay/no stay kind of stuff. We're in between T-two and T-three, and we use a cryogenic bottle, super critical helium, to pressurize our descent engine. Again, one of the things you can never test, the heat soak-back from the engine and the surface now is raising the pressure in that bottle very dramatically, and now we're wondering if this damned thing's going to explode and what the hell are we going to do about it. The fortunate thing was that they had designed some relief valves. They had a pressure disc in there. If the pressure got so high, it actually blows the disc and the valve, rather than blowing the bottle up. So we're all sweating this thing out here. We're trying to get everything re-synced for the next lift-off, and it's just time, which is almost two hours, between T-two and T-three stay/no stay, it just goes through incredibly quick. Throughout this whole period of time, except for the instant of hearing the cheering, you never got a chance to really think: 'We've landed on the Moon.'*

Afterwards Armstrong reflected upon his choice of words for the stepping out on the surface: 'That's one small step for man. One giant leap for mankind.'

> Armstrong: *I thought about it after landing, and because we had a lot of other things to do, it was not something that I really concentrated on but just something that was kind of passing around subliminally or in the background. But it, you know, was a pretty simple statement, talking about stepping off something. Why, it wasn't a very complex thing. It was what it was. I didn't want to be dumb, but it was contrived in a way, and I was guilty of that.*

An Embarrassing Failure

Whilst history was being made with *Apollo 11*, the Soviets were also trying to land *Luna 15* on the Moon in an attempt to upstage it. The responsibility for the *Luna 15* mission fell on the shoulders of First Deputy Minister of General Machine Building Georgi Tyulin, a 54-year-old retired artillery general. Tyulin ran into trouble with the spacecraft after only one day of flight. Controllers detected unusually high temperatures in the propellant tanks that would be used for take-off from the lunar surface after the collection of the lunar sample. After a quick analysis, a makeshift solution was proposed whereby the spacecraft would be turned in such a way that its suspect tank would lie in the Sun's shadow at all times.

Luna 15 fired its main engine to enter lunar orbit at 1 p.m. Moscow Time on 11 July, five days before *Apollo 11* took off. Its second orbit correction on 19 July would position the craft over its landing corridor. However, Soviet scientists did not anticipate the ruggedness of the lunar surface, and mission controllers spent three or four days finding a suitable landing site. Originally, their plan was to put down the lander less than two hours after *Apollo 11*'s touchdown, but the delays prevented that. Controllers finally commanded *Luna 15* to fire its descent engine at 6.47 p.m. Moscow Time on 21 July, a little more than two hours prior to Armstrong and Aldrin's planned lift-off from the Moon. They followed the signals from *Luna 15* as it descended. Six minutes before it was scheduled to land, all data suddenly ceased. Later analysis showed that the spacecraft

> ## 'That's one small step for man. One giant leap for mankind.'
>
> NEIL ARMSTRONG

had hit the side of a mountain. TASS announced that *Luna 15*'s research programme had been completed and the spacecraft had reached the Moon in the pre-set area. But even if *Luna 15* had worked perfectly and had returned with a soil sample it would have returned to Earth two hours and four minutes after the splashdown of *Apollo 11*. The race was over before it had begun.

The evening after the *Apollo 11* Moon landing someone placed a bouquet of flowers next to the grave of President John F. Kennedy at the Arlington National Cemetery. Attached was a note saying: 'Mr President. The Eagle has landed.'

'Houston – we've had a problem'

SUCCESS AND SUCCESSFUL FAILURE

APOLLO 12 AND APOLLO 13
1969–1970

The successful and almost flawless *Apollo 11* mission was followed in 1970 by a second manned Moon mission – *Apollo 12*, which experienced a lightning strike during take-off. However, this event now seems insignificant compared with the high drama that surrounded the next mission – *Apollo 13*. But thanks to the ingenuity of Mission Control and the calmness of those on board the stricken craft, disaster was averted and the crew returned safely to Earth.

On 11 October 1969 *Soyuz 6* thundered into orbit carrying two rookie cosmonauts, Georgi Shonin and Valeri Kubasov, both 33 years old. It had been almost ten months since the last Soviet manned flight. This was to be a shakedown mission to identify and iron out problems. Among its objectives were perfecting spacecraft control systems, testing navigational devices and taking photographs of the Earth.

Mounting rumours of more Soyuz launches were confirmed the following day, when *Soyuz 7* lifted off with more rookie cosmonauts, Anatoli Filipchenko, 41, Vladislav Volkov, 33, and Viktor Gorbatko, 34. Within two hours of launch, *Soyuz 8* was launched on 13 October 1969, with veteran cosmonauts Colonel Vladimir A. Shatalov and Aleksei Yeliseyev on board. By the following day the three spacecraft were in a common orbit. As planned, *Soyuz 7* and *Soyuz 8* approached each other to within a distance of 500 metres (1640 ft), while *Soyuz 6* watched from close by. Docking between *Soyuz 7* and *Soyuz 8* had been planned to be semiautomatic, with the Igla system bringing the two ships to a distance of 100 metres (328 ft).

That the mission was a complete mess was underlined in a US intelligence report, declassified in 1997. The five rendezvous attempts during the mission were all unsuccessful for various reasons. The first was unsuccessful because the automatic rendezvous system failed, and it was followed by a manual attempt that used too much fuel when trying to dock. Further attempts also failed. Despite this, the cosmonauts' return to Moscow was turned into a national celebration.

Charles 'Pete' Conrad examines the
unmanned *Surveyor 3* spaceprobe
during the *Apollo 12* mission.

Despite events behind the scenes, Soviet propaganda attempted to justify their space 'successes' and diminish the achievements of Apollo. The Americans were involved in a costly, empty race for political reasons, the Soviets stated, with no thought as to the good of the people or the real benefits of spaceflight. After Brezhnev finished his speech welcoming the Soyuz crews, it was clear to most in the Soviet space programme that the Moon race was over and a new era had begun – the age of the space station.

Apollo 12

The second Moon landing mission, *Apollo 12*, was launched into a thunderstorm on 14 November, but it was nearly over before it began – it was struck by lightning. Alan Bean was the Lunar Module pilot:

> *I did not know we were hit by lightning. I had no window. The window was covered then, and certainly, with all the noise and vibration, there's nothing. So many different caution, warning things came on. All the electrical system lights came on, every single one. This was no failure that we'd ever practised. In the backup for Apollo 9, in the flight for 12, we did every failure they had ten times. There was no failure that was even close. I looked at that control panel, and I thought: 'What could cause that? Everything there had gone.' So I said: 'We're getting ready to go into orbit without a Service Module.' I was thinking: 'What can I do about that?' So I'm there dividing my time between thinking: 'What is going wrong that would give us this indication?' And: 'Here's my chance, and I don't have the slightest idea what to do.' So my brain is not even able to completely concentrate on solving the problem. I was doing all that. Then I'd hear Pete and Dick over there. They're working on their part of the problem, and they start talking about lightning, and that doesn't mean a lot to me, because I'm still trying to figure out what to do.*
>
> *Then they call to get me to throw a switch, which I did, I didn't remember what the switch was for, either. I wasn't in any big hurry, because we were headed up to orbit, and I didn't want to screw that up by messing around over here. So I tried one and it stayed on. So I said: 'Wow. That doesn't even show.' You know, I checked the amps and volts. It worked good. It was great. I put on another one. It did the same thing; it worked great. I put on the third one. Each time I was waiting for something to go: 'Beep!' you know, and everything go off again. It never did.*

Dusty Moonwalks

Having survived the lightning strike, *Apollo 12* headed for the Moon. The Command Module pilot was Dick Gordon:

> *We were assigned a specific target called Surveyor 3 which was in a specific crater on the lunar surface. And the reason for learning how to do that was later missions we're going to land alongside 13,000 foot mountains and big valleys that were 600 metres deep and a kilometre across, the highlands of Descartes and Hadley Rille on 15 and Taurus-Littrow on 17. Later on, these sites were not picked but we knew that we wanted to go to other places that were going to require precise navigation.*

Lunar Module Intrepid landed in the Ocean of Storms, a region of the Moon that had been visited many times by unmanned spaceprobes; *Luna 5*, *Ranger 7* and *Surveyor 3* which had been on the Moon since April 1967. Intrepid touched down only 200 metres (656 ft) away. When he stepped onto the surface, Pete Conrad said: 'Whoopie! Man, that may have been a small one for Neil, but that's a long one for me.'

During two moonwalks lasting a total of seven hours and 45 minutes they collected rocks, set up experiments and walked over to *Surveyor 3*, removing several pieces from it to take back to Earth to see how it had fared during its two and a half years of exposure on the Moon. Only one thing marred the mission. To improve the quality of television pictures from the Moon over the ghostly black and white images returned from *Apollo 11*, a colour camera was included. Unfortunately, Al Bean accidentally pointed it at the Sun, burning it out. Situated on the crew's spacesuit cuffs were flipover checklists. The backup crew inserted reduced-sized pictures of *Playboy* centrefolds. Presumably a space first!

> Dick Gordon: *They had a good time on the lunar surface. They came back so damn filthy that I wouldn't let them in the Command Module. I made them strip, take every bit of clothes off they had. It had an extraordinary amount of dust that clung to their suits. When I looked into that Lunar Module when they took that hatch apart, all I could see was a black cloud in there, I didn't see them at all. I looked in there and said: 'Holy smoke. You're not getting in here and dirtying up my nice clean Command Module.' So they passed the rocks over, they took off their suits, passed those over, took off their underwear, and I said: 'Okay, you can come in now.'*

On the Brink of Catastrophe – *Apollo 13*

Alan Shepard, the first American in space, had overcome his medical problem and was due to command *Apollo 13*. But NASA was concerned that he would not be ready to fly *Apollo 13*, so they asked Jim Lovell, Fred Haise and Ken Mattingly – originally due to fly in *Apollo 14* – if they would take over the earlier mission and let Shepard train for the later one. Then Ken Mattingly was exposed to measles, to which he was not immune. Shortly before launch he was replaced by Jack Swigert. Launch was on 11 April 1970.

> Lovell: *We're out two days before the accident happens. But 30 hours after we took off, we got onto a different course because the course we were on originally was called a free-return course to allow us to get back to the Earth. But – about 30 hours, we changed course to land at this place – we were going to land at a place called Fra Mauro and the sunlight would be in the proper position to see the shadows. And then two days out, on this hybrid course, the explosion occurs. We just finished a TV programme. That was the last thing that evening. I think it was either 9 or 10 o'clock back here at Houston. And I'm coming back down through the tunnel, and suddenly there's a hiss-bang! And the spacecraft rocks back and forth. The lights come on and jets fire, and I looked at Haise to see if he knew what caused it. He had no idea. Looked at Jack Swigert. He had no idea. And then of course, things started to happen. The light came on. Something was wrong with the electrical system. We started – we eventually lost two fuel cells. We couldn't get them back. Then we saw our oxygen being depleted. One tank was completely gone. The other*

tank had started to go down. Then I looked out the window, and we saw gas escaping from the rear end of my spacecraft.

It was Jack that said: 'Houston – we've had a problem.' And Houston said: 'Say again, please?' And I say: 'Houston, we have a problem. We have a main B bus undervolt.'

Haise: *At the time of the explosion, I was in the Lunar Module. I was still buttoning up and putting away equipment from a TV show we had completed, and really we – subsequently we were going to get ready and go to sleep. I knew it was a real happening, and I knew it was not normal and serious at – just at that instant. I did not necessarily know that it was life-threatening. Obviously I didn't know what had caused it.*

Within a very short time, though, I had drifted up into the Command and Service Module to my normal position on the right, which encompasses a number of systems – the electrical system, cryogenics, fuel cells, communication, environmental systems – and I was just looking at the array of warning lights. It was confusion in my mind because we had never had a single credible failure that would have caused that number of lights on at one time.

One thing, though, just looking over the instrument panel that became very clear in short order was the fact that the pressure meter, the temperature, and the quantity meter needles for one of the oxygen tanks was down in the bottom of their gauges. These are different sensors, so it was unlikely that this was false. So it effectively told me we had lost one oxygen tank. My emotions at that time went to just a sick feeling in the pit of my stomach, because I knew by mission rules, without reference, that that meant the cancellation of the lunar mission. We were in an abort mode but still not life-threatening, because we had a second oxygen tank, I thought, which looked to be still there. And we'd have stayed fully powered-up and then just took an abort mode to come back home in a – with everything fully powered. It took some minutes to become obvious that there was, for whatever reason – that there was a leak that the explosion had caused in the second oxygen tank. Either the tank or one of the lines. And – but a small leak. And when that – when it became obvious it was dwindling or losing oxygen, then the handwriting was on the wall that the Command Module, it was going to die and have to be powered-down.

Lovell: *The thought crossed our mind that we were in deep trouble. But we never dwelled on it. We never, you know, sort of gave up and said: 'What are – what's going to happen if we don't get back? Where are we going to be?' My thoughts were this: if everything failed and we still had life-support in the Lunar Module but we couldn't get back to the Earth, you know, the heat shield was damaged or we just went past the Earth. I said that: 'We will send back information. We'll keep on operating as long as we can. And then, that's the end of the deal.' So, that was what I had planned to do in my mind should – you know, should something happen. People often ask, you know, this poison pill deal's ridiculous.'*

The critically damaged *Apollo 13* Service Module photographed by the crew after jettisoning. An entire panel is missing, blown away by an explosion in oxygen tank number two.

Mission Controller Gene Kranz remembers the dramatic events of the night well:

> They'd go into the Lunar Module, and they also had a television broadcast of the Lunar Module. The television broadcast was concluded. And the final – we were in the process of closing out the items in the shift prior to hand over to Glynn Lunney's Black Team. After television broadcast was concluded, the wives and families had been behind me in the viewing room, and as they left we sort of waved: 'Okay,' etc., 'Adios,' and they went off. They turned the lights out in the viewing room behind me, and the final thing we had to do was to get the crew to sleep. And we have a very detailed pre-sleep checklist we'd go through. It's about five pages in length. And we had gone through each one of these checklist items very meticulously because in Mission Control, the greatest error that always lends to a lot of levity at the post- mission party is for some flight controller to miss something in this pre-sleep checklist that cause us to wake up the crew. And we have a series of awards we give out at the parties if this happens. And it's not all the jollies you get; you get really ridden pretty hard. So, we were very meticulously following through this checklist. We were down to the final item in the checklist. We were getting ready to close it out.
>
> And we were down to the final entry, and – the cryogenics, the fuels that we use on board the spacecraft, are oxygen and hydrogen. It's a super dense, super cold liquid at launch at temperatures of -300 to -400 °F, packed in vacuum tanks. But by the time you're two days into the mission, you've used some of these resources. And these consumables have turned into a very thick, soupy fog or a vapour in the tank. And like fog on Earth, it tends to stratify or develop in layers. So, inside the tanks, we have some fans we turn on to stir up this mixture and make it uniform so we can measure it. Then we use some heaters to raise the pressure for the sleep period. Well, we had asked the crew to do this. In the meantime, the next control team was reporting in for shift handover, so the noise level in the room was building up; and their flight director, Glynn Lunney (he was the leader of the Black Team, and we used colours to identify those teams), was sitting next to me at the console. He was reading my flight director's log. And we advised the crew that we wanted a cryo stir. Jack Swigert acknowledged our request, and he looked behind him and coming through the tunnel, from the Lunar Module, was Fred Haise. Sy Liebergot at this time had the responsibilities for the cryo systems, had now switched his attention to the current measurements that he had. And Swigert started the cryo stir.

1969

14 November US launches *Apollo 12* successfully to the Moon

1970

11 February Japan launches its first satellite, making it the fourth nation with a space rocket powerful enough to launch satellites into Earth orbit, after the USSR, the US and France

11 April *Apollo 13* blasts off on a mission to the Moon

13 April Four-fifths of the way to the Moon, *Apollo 13* is crippled when a tank containing liquid oxygen bursts. Despite this, the three-man crew manage to return to Earth

17 April *Apollo 13* crew splash down safely in the Pacific Ocean

24 April China launches its first satellite

Liebergot saw the currents increase indicating the stir had started. All of a sudden I get a series of calls from my controllers. My first one is from guidance. It says: 'Flight, we've had a computer restart.' The second controller says: 'Antenna switch.' The third controller says: 'Main bus undervolt.' And then from the spacecraft I hear: 'Hey, Houston, we've had a problem." And there was a pause for about 5 seconds. And then Lovell comes onboard to say: 'Hey, Houston, we've got a problem.' Within Mission Control, literally nothing made sense in those first few seconds because the controllers' data had gone static briefly; and then it – when it was restored, many of the parameters just didn't indicate anything that we had ever seen before. Down in the propulsion area, my controllers all of a sudden saw a lot of jet activity. Jets were firing. We then see Lovell – and this is all happening in seconds – we then see Lovell take control of the spacecraft and fly into an attitude so he can keep communicating with us.

And for about 60 seconds, literally, the calls kept – I mean, just coming in. But they made no sense. They made no pattern, right on down the line, until finally the training that's given the controllers kicked in. And very meticulously, they started making the calls that were called – relayed up by Jack Lousma, my Capcom at that time.

For probably about 60 to 90 seconds, it's literally chaos in this place. And then it's amazing how this whole thing, it starts to take focus. We still don't have the slightest clue what's going on. But instead of listening to every crew call and – controller call and relaying it up, I start being much more selective in this process. Because I'm starting to get the feeling that this isn't a communications glitch. I'm about five minutes into this problem right now. It's something else. We don't understand it. So, we proceed very meticulously. And I call the controllers up and I tell them that: 'Okay, all you guys, quit your guessing. Let's start working this problem.' Then I use some words that sort of surprised me after the fact. I say: 'We've got a good main bus A. Don't do anything to screw it up. And the Lunar Module's attached, and we can use that as a lifeboat if we need to. Now get me some backup people in here and get me more computing and communications resources.' I'd said these words, but then I immediately went back to tracking this thing.

The second oxygen tank is now continuing to decrease. Two of our fuel cells are off line, and these are our principal power-generation systems that we use. Liebergot then comes to me and says: 'Hey, flight, I want to shut down fuel cells one and three.' And I say: 'Sy, let's think about this.' And he says: 'No, flight, I think that's the only thing that's going to stop the leaks.' And then I go back to him the third time and I say: 'Sy,' and he says: 'Yeah, flight, it's time for a final option.' And very reluctantly I agree to advise the crew that we're going to shut down fuel cells one and three. I think this is probably the point in the mission where everybody has realized that we've now moved into a survival mode because with two of the three fuel cells shut down, we're not going to the Moon any more. We're going to just be damn lucky to get home alive.

By this time, Lovell's called down and indicating they're venting something. And we've come to the conclusion that we had some type of an explosion onboard the spacecraft; and our job now is to start an orderly evacuation from the Command Module into the Lunar Module. At the same time, I'm faced with a series of decisions that are all irreversible. At the time the explosion occurred, we're about 200,000 miles from Earth, about 50,000 miles from the surface of the Moon. We're entering the phase of the mission – we use the term 'entering the lunar sphere of influence'. And this is where the Moon's gravity is becoming much stronger than the Earth's gravity. And during this period, for a very short time, you have two abort options: one which will take you around the front side of the Moon, and one which will take you all the way around the Moon.

If I would execute what we call a 'direct abort' in the next two hours, we could be home in about 32 hours. But we would have to do two things: we'd have to jettison the Lunar Module, which I'm thinking of using as a lifeboat, and we'd have to use the main engine. And we still have no clue what happened onboard the spacecraft. The other option: we've got to go around the Moon; and it's going to take about five days but I've only got two days of electrical power. So, we're now at the point of making the decision: which path are we going to take? My gut feeling, and that's all I've got, says: 'Don't use the main engine and don't jettison this Lunar Module.' And that's all I've got is a gut feeling. And it's based, I don't know – in the flight control business, the flight director business, you develop some street smarts. And I think every controller has felt this at one time or another. And I talked briefly to Lunney, and he's got the same feeling.

Then John Aaron said: 'There's no way we're going to make five days with the power in the Lunar Module. We got to cut it down to at least four days, maybe three.' So, we were now moving ahead. The team split up and moving in several different directions. I had one team working power profiles. I had another group of people that was working navigation techniques. I had a third one that was integrating all the pieces we need. My team picked up the responsibility to figure out a day to – a way to cut a day off the return trip time.

During Apollo 9, we did a lot of testing of the Lunar Module engine while the two spacecraft were docked together. And immediately as soon as we recognized we had to perform a manoeuvre to speed up our return journey, that's the set of procedures we fell back to. We updated these procedures, based on the situation at hand. My team came back on console and executed these procedures, and increased our velocity on return by almost 1000 feet per second. Changed the landing point from the Indian Ocean now to the South Pacific. We sent the aircraft carrier Iwo Jima to the new landing location.

Apollo 13 was now in survival mode. Everything on the spacecraft had to be conserved for re-entry. All the power they had was about the equivalent of a 200 watt light bulb. Everything else had to be powered down. Another problem occurred: the lithium hydroxide canisters that removed harmful carbon dioxide (CO_2) from the air were in the shut down Command Module and the canisters in the Lunar Module were running out. The problem was that they were not interchangeable. The canisters in the one wouldn't fit in the other.

Mattingly: *We had another consumable. I don't remember anybody forecasting that we would have a CO_2 problem, but as soon as the light came on, we recognized it. In the movie [Apollo 13, starring Tom Hanks] there's this really neat scene where they've got a tableful of stuff and Robert Smylie dumps a bag of what they had in space on the table, and he says: 'Figure something out.' Well, the real world is better than that. The real world is that we had had a simulation, and I think it was on Apollo 8, I believe, where the simulation supervisor had jammed one of the cabin fans with a screw that floated loose. I think they had broken some electrical connections or done something of that ilk. The conclusion, you know, the simulations were done with the rule that the simulation may be four hours, but it's not over until everything is under control. So sometimes those things got to be rather lengthy simulations. The solution that they came up with was that they could make a way to use the vacuum cleaner in the Command Module with some plastic bags cut up and taped to the lithium hydroxide cartridges and blow through it with a vacuum cleaner. So, having discovered it, they said: 'Okay, it's time for beer.' Well, on 13, someone says: 'You remember what we did on that sim? Who did that?' So in nothing short, Joe Kerwin showed up, and we talked about: 'How did you build that bag and what did you do?' Oh, it was easy. Solving that problem took an hour, maybe two. Because it's real now, they made him build a demonstration model, so that took another 30 minutes. Then, 'How are we going to tell these guys in the cockpit?' And the answer was, if you just said go tape your lithium canisters to the suit hose, that's probably all they had to say.*

Haise: *The vehicle had gotten very cold. We were a little warmer than freezing but not a lot. And that kind of wore on you after a while. We did not have adequate clothing to handle that situation. We did put on every pair of underwear we had in the vehicle. Jim Lovell and I wore our lunar boots, the boots we would normally put over our spacesuit boots on the lunar surface.*

Due to the cold there were fears about what the extensive condensation would do to the controls in the Command Module when it was time to power it up before re-entry. Fortunately it worked. Many believe that was due to the extensive modifications made after the *Apollo 1* fire.

Safely Back Home

Prior to re-entry they jettisoned the Lunar and Service Modules. One side of the Service Module had been ripped out by the explosion. The last task was re-entry. The Command Module entered the atmosphere.

Kranz waited: *There isn't any noise in here. You hear the electronics. You hear the hum of the air conditioning occasionally. In those days, we used to smoke a lot. Somebody would only hear the rasp of the Zippo lighter as somebody lights up a cigarette. And you'd drink the final cold coffee and stale soda that's been there. And every eye is on the clock in the wall, counting down to zero. And when it hits zero, I tell Kerwin: 'Okay, Joe, give them a call.' And we didn't hear from the crew after the first call. And we called again. And we called again.*

And we're now a minute since we should've heard from the crew. And for the first time in this mission, there is the first little bit of doubt that's coming into this room that something happened and the crew didn't make it. But in our business, hope's eternal, and trust in the spacecraft and each other is eternal. So, we keep going. And every time we call the crew, it's: 'Will you please answer us?' And we were one minute and 27 seconds since we should've heard from the crew before we finally get a call. And a downrange aircraft has heard from the crew as they arrive for acquisition of signal. And then almost instantaneously from the aircraft carrier, we get: 'A sonic boom, Iwo Jima. Radar contact, Iwo Jima.' And then we have the 10-by-10 television view. And you see the spacecraft under these three red and white parachutes, and the intensity of this emotional release is so great that I think every controller is silently crying. You just hear a 'Whoop!'

Back on Earth they re-enacted the drama. Fred Haise recalled what happened:

It was interesting that when Jim Lovell and I, after the flight, went just out of curiosity – went back into a Lunar Module simulator, we could not replicate the time of that activation in … the nice, calm conditions of a ground simulator that we had done in flight.

Lovell: *NASA will claim that they are absolutely not superstitious. But I'll bet you my last dollar, they'll never name another spacecraft 13.*

Spurred on by the *Apollo 13* setback the Soviets tested their own version of the lunar lander in Earth orbit in 1970 and 1971, but there was nothing to be gained by it. It took them a while to realize it.

> ## 'NASA will claim that they are absolutely not superstitious. But I'll bet you my last dollar, they'll never name another spacecraft 13.'
>
> JIM LOVELL

'I cried a little'

THE RETURN OF ALAN SHEPARD

SOYUZ 9, APOLLO 14 AND SALYUT 1
1970–1971

Alan Shepard, the first American in space and a veteran of the Mercury and Gemini programmes of the early 1960s, feared that his opportunity to fly in space again had been dashed by illness. However, he not only overcame this problem but also found himself in charge of a mission to the Moon – *Apollo 14*. The start of the 1970s also saw the Russians launch the first-ever space station.

The last of the pre-space station Soyuz craft was *Soyuz 9*. Its crew was 40-year-old Colonel Andriyan Nikolayev as the commander and 34-year-old civilian Vitally Sevastyanov as the flight engineer. It lifted off on 1 June 1970. For Nikolayev, it was his second spaceflight, having flown in space eight years before in 1962 as the pilot of *Vostok 3*. At the time Neil Armstrong was on an official visit to the Soviet Union. On the night of the launch, at the Cosmonaut Training Centre near Moscow, he was clearly surprised when his host, cosmonaut Major General Beregovoi, turned on the television to view film of the *Soyuz 9* launch, telling Armstrong: 'This is in your honour.'

Soyuz 9 stayed in space for 17 days, exceeding the record set by *Gemini 7* in 1965, but the crew displayed the first real signs of fatigue and decrease in working efficiency on their 12th day in orbit. Kamanin wrote in his diary: 'Nikolayev and Sevastyanov look somewhat puffy, and listlessness and irritability can be sensed in their actions.' After landing, ground crews reached the cosmonauts and found that they were unable to get out of the ship themselves and had to be carried out. It was decided to cancel the flight of the crew to Moscow Airport. Instead, they rested for a day.

> Kamanin: *When I entered the aircraft's cabin, Sevastyanov was sitting on the sofa. while Nikolayev was at a small table. I knew they were having a hard time enduring the return to the ground, but I had not counted on seeing them in such a sorry state. Pale, puffy, apathetic, without the spark of vitality in their eyes – they gave the impression of completely emaciated, sick people.*

Apollo 14

The US space effort did not recover from the *Apollo 13* accident until January 1971. The cause was found to be relatively simple – a manufacturing fault in one of the liquid oxygen tanks. Alan Shepard, grounded after his Mercury flight due to a medical problem, returned to space and blasted off for the Moon in *Apollo 14*, having featured in one of the most remarkable stories in spaceflight.

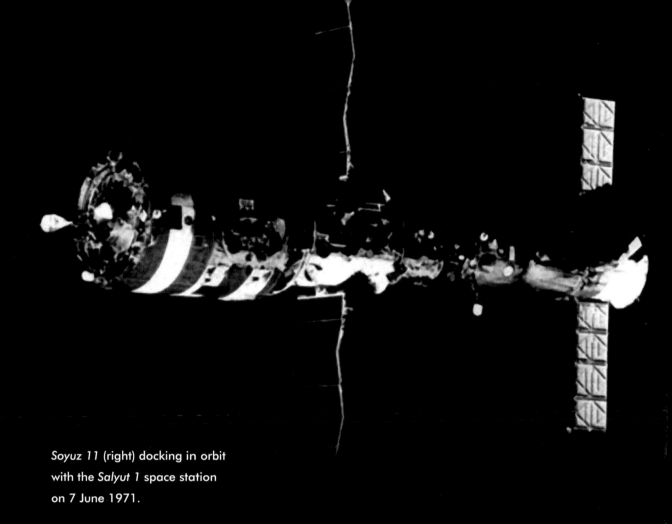

Soyuz 11 (right) docking in orbit
with the *Salyut 1* space station
on 7 June 1971.

Alan Shepard had been suffering from Menière's disease that causes elevated pressure in the inner ear. After NASA grounded him he contemplated his options:

> So there I was, what do I do now? Do I go back to the navy? Do I stick around with the space programme? What do I do? I finally decided that I would stay with NASA and see if there wasn't some way that we could correct this ear problem. Several years went by, there was some medication which alleviated it, but I still couldn't fly solo. Can you imagine the world's greatest test pilot has to have some young guy in the back flying along with you? I mean, talk about embarrassing situations!
>
> It was Tom Stafford who came to me and said he had a friend in Los Angeles who was experimenting correcting this Menière's problem surgically. And so I said: 'Gosh that's great. I'll go out and see him.' So he set it up. I went on out there. The fellow said: 'Yeah, we do. What we do is we make a little opening there, put a tube in so that it enlarges the chamber that takes that fluid pressure, and in some cases it's worked.' And I said: 'Well, what if it doesn't work?' And he said: 'Well, you won't be any worse off than you are, except you might lose your hearing. But other than that …' So I went out there under an assumed name, Victor Poulos. The doctor knew and the nurse knew who I was but nobody else knew … So, Victor Poulos checks in and they run

1970

1 June Russia launches *Soyuz 9* during a visit to the Soviet Union by US astronaut Neil Armstrong

22 July Russian *Venera 7* lander transmits data from the surface of Venus

1971

4 February *Apollo 14* lander Antares descends to the Moon surface with Alan Shepard and Ed Mitchell on board

19 April Russians launch *Salyut 1*, the first orbital space station

the operation … it's not that traumatic, obviously, because after about a day I was out of there. Of course it was obvious when you look at the big ball of stuff over my ear when I get back home. But NASA started looking at me. And several months, several months, several months went by, and finally they said: 'Yes, all the tests show that you no longer are affected by this Menière's disease.' So there I was, having made the right decision.

When NASA finally said I could fly again, I went to Deke Slayton who was in charge of astronaut assignments and said: 'We have not announced publicly the crew assignment for Apollo 13. I have a recommendation to make.' I had picked two bright, young guys – one of them a Ph.D, and one of them a heck of a lot smarter than I was – and made up a team to go for an Apollo flight. I said: 'I would like to recommend that I get Apollo 13, with Stu Roosa as Command Module and Ed Mitchell as lunar pilot.' Deke said: 'I don't know. Let's try it out.' So we sent it to Washington, and they said: 'No, no way.' So we said: 'Now wait a minute. Shepard's got to be at least as smart as the rest of these guys, maybe a little smarter.' And they said: 'Well, we know that. But it's a real public relations problem. Here this guy's just gotten un-grounded and all of a sudden, boom! He gets premier flight assignment.' So then the discussion went on for several days and finally they said: 'All right, we'll make a deal. We'll let Shepard have Apollo 14. Give us another crew for Apollo 13,' and so that's what happened.

The flight had gone extremely well. We'd had one or two docking problems earlier, a problem with something floating around in the abort switch, which closed as if we were pushing the abort switch closed. All of these were taken care of. Now we're on our way down, flying up on our backs with the engine pointing that way, slowing down, and getting gradually more steeper and more steeper. We had a ruling that the computer had to be updated by the landing radar; reason being is that while you're on your back obviously you can't see the ground, you can't see the mountains, you can't see the rocks, or anything. So we had a rule that said if the landing radar was not updating the computer by the time you were down at a level of about 13,000 feet, then you have to abort; you have to get out of there. Well, the landing radar wasn't working. They called us up and said: 'Your landing radar is not working.' We said: 'Thank you very much, we're aware of that.' And then a little bit further on they said: 'You know what the ground rule is about aborting if you're not at 13,000 feet.' Well, yeah we knew that. Finally some bright young man over in the control centre said: 'Hey your landing radar is working, but it's locked up on infinity. Have them pull the switch, reset it, and see if it works.' So we pulled the circuit breaker, put it back in, and sure enough the landing radar came on. And shortly after that we got cleared to land with a margin of 1000 feet or so, which was a close thing. As soon as we pitched over there was beautiful Fra Mauro, just the way I had seen it hundreds of times from the scale model. We came on down, made a very, very soft landing. As a matter of fact soft enough so that even though we'd landed in a slight crater, the uphill leg didn't crush like it was supposed to. We had crushable material in the lining. It was a slight ring wing down perfect landing. We shut off the switches and Ed Mitchell turned to me and said: 'Alan, what were you going to do if the landing radar had not been working by 13,000 feet?' I looked at him and I said: 'Ed, you'll never know.'

Of course the first feeling was one of a tremendous sense of accomplishment, I guess if you will. A tremendous sense of realizing that, 'Hey, not too long ago I was grounded. Now I'm on the Moon.' There was that sense of self-satisfaction immediately. But then that went away, because we had a lot of work to do. But I'll never forget that moment. Another moment which I will never forget is after Ed had followed me down and we had set out some of our equipment, taken the emergency samples, we had a few moments to look around, to look up in the black sky – a totally black sky, even though the Sun is shining on the surface it's not reflected, there's no diffusion, no reflection – a totally black sky and seeing another planet: planet Earth. Now planet Earth is only four times as large as the Moon, so you can really still put your thumb and your forefinger around it at that distance. So it makes it look beautiful; it makes it look lonely; it makes it look fragile. You think to yourself, just imagine that millions of people are living on that planet and don't realise how fragile it is. I think this is a feeling everyone has had and expressed it in one fashion or another. That was an overwhelming feeling in seeing the beauty of the planet on the one hand but the fragility of it on the other. Then I cried a little.

The First Space Station

The first of the USSR's space stations was launched on 19 April 1971 by a *Proton* rocket. Called *Salyut 1*, it was the first of a series of nine single-module space stations. They were launched over a period of 11 years to 1982 to investigate the techniques and problems of lengthy stays in space as well as to perform a wide variety of experiments in the microgravity conditions of orbit. In a way the single-core Salyut stations were stepping stones, allowing space technology to develop from engineering development modules to larger, more complicated space stations designed for long-duration occupancy. Ultimately they paved the way for the *Mir* space station and the *International Space Station* currently in orbit.

But whatever its technological legacy would eventually become, *Salyut 1* was the scene of the worst ever tragedy in orbit.

Ten to 15 seconds of agonizing consciousness

DEATH IN SPACE

SOYUZ 10 AND 11, SALYUT 1 AND 2
1971 AND 1973

The Soviets' attempts to successfully rendezvous their Soyuz spacecraft with the orbiting Salyut space station as a prelude to landing on the Moon were proving to be disastrous. Finally, a tragic accident during the re-entry of *Soyuz 11* heralded the cancellation of their lunar landing ambitions.

Soyuz 10 was launched a few days after *Salyut 1* – the Soviet's first space station. It had a crew of three. Vladimir Shatalov and Alexei Yeliseyev had flown together in *Soyuz 4, 5* and *8*. Nikolai Rukavishnikov was on his first flight. Meanwhile, the *Salyut 1* space station had encountered problems. After it had reached orbit, controllers realized that the large exterior cover protecting its telescope had not been jettisoned. It was a major blow, and it was not to be the only problem.

Despite a successful launch, the outlook for the *Soyuz 10* mission was not good. As well as the problem with the cover and other failures, six of the eight ventilation units in the life-support system had failed, raising the prospect of high levels of carbon dioxide inside the space station. At a distance of 10 miles (16 km) from the station Shatalov switched on the Igla rendezvous docking system, which successfully brought *Soyuz 5* to within 200 metres (656 ft) of *Salyut 1* before he took over manual control to dock. However, the docking indicator warning light suggested that hard-docking had not taken place. Ground telemetry confirmed that full docking had not occurred and that there was still a 9-centimetre (3.5-in) gap between the two vehicles. Shatalov attempted to push the two ships together by firing the *Soyuz* engines, but it did not work. After four orbits he was ordered to undock, but for some reason the two craft would not separate. Suddenly they were in severe danger. There were two options. One was to dismantle the docking apparatus, detach it from *Soyuz 5* and move away from the station. The other was to separate the spheroid living compartment from the *Soyuz 5* spacecraft, leaving the other compartment docked to *Salyut 1*. In both cases, the station would be unusable in the future because its only docking port would be occupied. While mission controllers contemplated these unappealing scenarios Shatalov tried again to separate from *Salyut*, and to everyone's relief succeeded. *Soyuz 5* could not remain in space after this near-disaster, however; they had to plan for an emergency re-entry, performing the first-ever night landing in

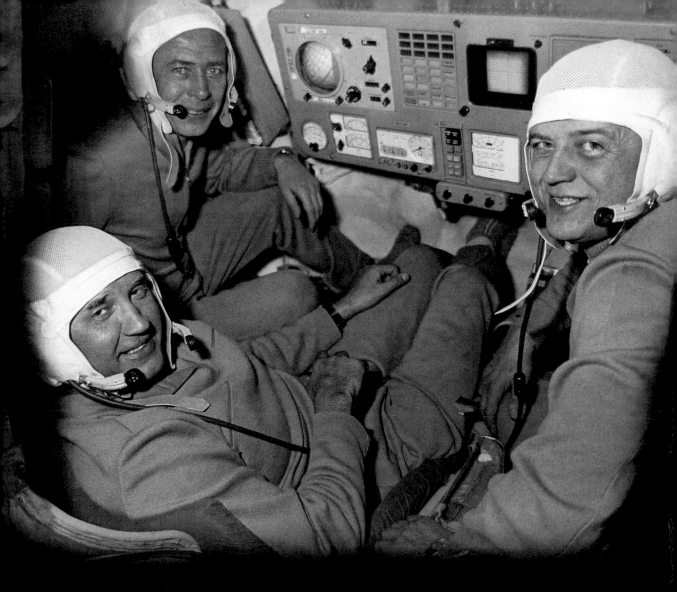

The doomed *Soyuz 11* crew (from left to right):
Georgi Dobrovolsky, Viktor Patsayev and Vladislav Volkov.

the Soviet programme. Upon re-entry the capsule filled with toxic fumes, causing Rukavishnikov to pass out. In total the aborted mission had lasted only one day, 23 hours, 46 minutes and 54 seconds. The Soviet media called it a success.

First Occupants of the Space Station

Two months later the Russians tried again with *Soyuz 11*. Originally the crew was to have been Leonov, Kubasov and Kolodin, but doctors suspected that Kubasov was about to contract tuberculosis. Under Soviet mission rules this meant the entire crew had to be replaced by their backups – in this case, Georgi Dobrovolsky, Vladislav Volkov and Viktor Patsayev. A reporter for the newspaper *Pravda* was at the Cosmonaut Hotel in Baikonur when the crew change took place. He later said that Leonov broke down

1971

23 April *Soyuz 10* is launched, intended to be the first craft to visit Soviet space station *Salyut 1*. But it experiences potentially disastrous docking problems and the mission is aborted

19 May American *Mars 3* orbiter and lander are launched successfully

6 June *Soyuz 11* is launched and docks successfully with *Salyut 1*. The crew suffer a catastrophic loss of air on the return journey and are dead on landing

third primary crew member, arrived at the hotel completely drunk. According to one report, Kolodin told Mishin that 'history would never forgive him' for his decision to send the backup crew.

The new crew left Earth on 6 June and docked successfully, becoming the first to occupy a space station. As the days turned into weeks they managed to carry out a full experimental programme, despite *Salyut 1*'s nagging problems. At last it seemed that the USSR had a real space success to celebrate. Reports from the cosmonauts in *Salyut 1* were shown on Soviet television. Many of their exchanges were humorous in nature. Soon they had become household names in the Soviet Union. Perhaps now they could start to put Apollo behind them?

The crew began their preparations to return to Earth after 23 days, having broken the space endurance record. It was clear that they were very tired. On the eve of 29 June they moved into *Soyuz 11*, which was docked to *Salyut 1*, and entered the descent capsule, sealing the hatch between it and the spherical living compartment. There was a major crisis at this point when the hatch open indicator light between the *Soyuz 5* living compartment and the descent apparatus failed to turn off. Tired and anxious, Volkov asked mission control what to do. They resealed the hatch, but the warning light remained on. Amazingly, ground control eventually recommended they cover the hatch sensor with a piece of paper, thus disabling it! Fortunately, the subsequent pressurization checks were satisfactory. The crew undocked and loitered for a little while taking exterior photographs of *Salyut 5*. They then started the process for automatic re-entry and routinely signed off communications. The *Soyuz* drifted out of voice contact. The main engine was to begin firing at 0135 hours and 24 seconds Moscow Time on 30 June, followed by separation of the three *Soyuz 5* modules.

An Agonizing End

Later, after hearing nothing from the returning crew, rescuers overflew the planned landing site. The capsule had landed safely, but the crew were dead inside. There is a heartbreaking video that emerged in Russia many years later showing the ground crew pulling the cosmonauts out of the capsule and attempting mouth-to-mouth resuscitation and heart massage. Investigations showed there was blood in the crew's lungs, nitrogen in their blood and haemorrhages in their brains – they had asphyxiated when the air escaped from their capsule. The crew, realizing they were losing air, had tried desperately to find the source of the leak – it was a valve that had been jolted open during re-entry as the Descent Module separated from the Service Module. They knew the danger and what would be the consequence if they did not rectify it.

The shoulder straps of all the cosmonauts were unfastened, and Dobrovolsky himself had become tangled in his straps as he struggled for freedom. Records indicated that his breathing rate shot up from a normal 16 breaths to 48 breaths per minute. They had ten to 15 seconds of agonizing consciousness – too little to do anything about the situation – and in less than a minute they were dead. The problematic valve was inaccessible behind their seats; ironically, a finger or an adhesive plaster over it would have saved them.

Afterwards, Kubasov's lung problem, which had effectively saved his life as well as those of Leonov and Kolodin, turned out to be only an allergic reaction. It would be over two years before a cosmonaut would venture into space again. *Salyut 1* re-entered and was destroyed in the Earth's atmosphere four months later.

The Soviets tried to resurrect their *N1* programme with another test launch on 27 June, but it only flew for 51 seconds. A year later they regrouped and tried again. This time the *N1* lasted 107 seconds before it broke up. The Moon project was finally cancelled in 1974. The mighty *N1s* were scrapped for use as sheds and shelters at Baikonur and, as was the habit in Soviet Russia, an attempt was made to obliterate them from history.

Salyut 2 was launched on 4 April 1973. It was not really a part of the Salyut series although it was included under that classification. In reality it was a prototype military space station called *Almaz*, but it malfunctioned in orbit and re-entered a few weeks later. The 'true *Salyut 2*' was launched a little later under the name *Cosmos 557*, but it did not work either and lasted only a week.

'Their majesty
overwhelmed me'

BRINGING BACK MOONBEAMS

APOLLO 15, 16 AND 17
1971–1972

The Apollo missions *15, 16* and *17* were notable not only for the well-executed, almost routine, way in which the journeys to and from the Moon were conducted. These three missions also carried lunar rover vehicles, which enabled the astronauts to explore terrain at some distance from the lander. Furthermore, the scientific purpose of the missions was emphasized by the fact that among the crew were trained geologists.

In July 1971 *Apollo 15* hit its target precisely – a large amphitheatre girded by mountains and a deep canyon on the eastern edge of a vast plain. Later, Commander David Scott said he would never forget the Command Module Endeavour hurtling through the Moon's strange nightscape. Above were the stars, below lay the Moon's far side, an arc of impenetrable darkness that blotted out the sky. As the moment of sunrise approached, barely discernible streamers of light – actually the glowing gases of the solar corona around the Sun – played above the Moon's horizon. Finally, the Sun exploded into view. In less than a second its harsh light flooded Endeavour, dazzling Scott's eyes.

The early lunar morning stretched into the distance. Long, angular shadows accentuated every hill and crater. As the Sun rose higher the moonscape turned the colour of gunmetal. The shadows shrank, and at lunar noontide it shone down on a bleached and featureless world.

The Lunar Geology

Dave Scott and Jim Irwin spent 67 hours on the Moon, landing in the bright morning of the 710-hour lunar day. Opening the top hatch, Scott made a preliminary survey of a world he described as still being in the epoch of its creation. Craters left by 'recent' meteorites millions of years ago stood out startlingly white against the soft beige of the gently undulating terrain. On the Moon the horizon is only 1.2 miles (2 km) away. To the south of Falcon, the Lunar Module, a 3300-metre (10,824-ft) ridge rose over the plain. To the east was an even higher summit. To the west was the Hadley Rille, snaking across the landscape and 300 metres (984 ft) deep. To the northeast was a great mountain towering 5000 metres (16,494 ft) above them. 'Their majesty overwhelmed me,' said Scott.

Scott was a space veteran on his third flight, but he was also a new breed of astronaut. Eight years of training in lunar geology had made him aware of intriguing details in the landscape and the rocks. A dark line, like a bathtub ring, smudged the base of the mountains. Was it left by the subsiding lake of lava that once filled the immense cavity of Palus Putredinis on the fringes of the vast Mare Imbrium (Sea of Rains) billions of years ago?.

Once on the lunar surface, Scott found the one-sixth Earth gravity more enjoyable than weightlessness, for while it enabled him to enjoy the almost same sense of buoyancy it still provided a reassuringly fixed awareness of 'up' and 'down'. Scott felt like an intruder, he said, in an eternal wilderness. The flowing moonscape reminded him of the Earth's uplands after a covering of snow. Most of the scattered rocks shared the same greyness as the dust around them, but he found two that were jet black, two that were pastel green, several with sparkling crystals, some coated with glass and one that was white. No wind blew, no sound echoed, only shadows moved.

Apollo 15 astronaut Jim Irwin loads the lunar rover prior to the first lunar extravehicular activity (EVA).

Acclimatizing to the Moon

At first, Dave Scott and Jim Irwin experienced a troubling deception in perspective. There were no trees, clouds or haze to determine whether an object was far away or near. Each of the three spacewalks was due to last seven hours, and they dug and drilled, gathered rocks and took photographs. Back in Falcon between excursions it took them two hours to remove their suits and do housekeeping chores. For the first 20 minutes or so, before the air filtration system purified the air, they were aware of a smell like that of gunpowder. But even when the smell disappeared, the moondust stuck to everything. To help them sleep they put shades over the windows.

By the third moonwalk they felt at home. Using the lunar rover on its first mission they ventured over the horizon – the first astronauts ever to do so. In case the navigation system on the lunar rover failed, Scott made a small cardboard compass. Although shrivelled in the savage lunar sunlight and covered with moondust, it would give him the bearings back to Falcon if he needed them. On their way back the astronauts even dared to take a short cut as the rover bounced between undulations and crater walls that hid their view of the Lunar Module for long minutes. Too soon they would have to leave the Moon, and already felt a sense of impending loss. They would never return to the plains of Hadley. Clutching the ladder, Scott raised his eyes from the moonscape and saw the vivid blue sphere of the Earth.

In times to come, other astronauts from Earth, or perhaps from elsewhere, would come this way and gaze upon the abandoned gear of *Apollo 15*. As with all Apollo landers, a plaque of aluminium fitted to the descent stage of the Lunar Module portrayed the two hemispheres of Earth, the name of the spacecraft, the date of the mission and the roster of the crew. The crew of *Apollo 15* left behind a falcon's feather and a four-leaf clover. In a little hollow in the moondust they placed a stylized figure of a man in a spacesuit alongside a metal plaque bearing the names of the 14 Russian and American spacemen who had given their lives so that man might explore the cosmos. Alongside, Scott placed a single book, the *Bible*.

Apollo 16 Leaves Its Mark

In April 1972 Charles Duke, the astronaut who communicated with Armstrong and Aldrin as they landed on the Moon, got his chance to land on the Moon himself on *Apollo 16*. When they reached lunar orbit, a malfunction in the Service Module controlling the angle of the booster nozzle nearly caused the cancellation of the lunar landing. Duke and Commander John Young were already undocked in their Lunar Module when their descent was postponed. Eventually it was decided that the problem was not a showstopper, and they got the go to descend. A communications problem meant that they had to start their descent with the windows facing out into space instead of towards the Moon, and needed to rely on the landing radar to update them on their altitude.

1971

30 July US *Apollo 15* astronauts Dave Scott and Jim Irwin land at Mare Imbrium on the Moon

1972

21 April *Apollo 16* astronauts John Young and Charles Duke explore the surface of the Moon with a lunar rover vehicle

7 December *Apollo 17*, the last American manned Moon mission to date, blasts off. The mission is the longest-ever lunar exploration (75 hours), and includes driving the lunar rover for about 22 miles (36 km)

Duke: *So at 7000 feet, the guidance programme manoeuvred the vehicle to windows forward down, and I saw the lunar surface close up for the very first time. It looked exactly like the mockup. 'John, there it is! You know, there's Gator. There's Lone Star and North Ray Crater.'*

We landed in the Descartes highlands of the Moon – a valley 8 to 10 miles across, and the objective was to explore to the south to a place we called Stone Mountain and then to the north, 3 or 4 miles, to a place called North Ray Crater, which was at the base of the Smoky Mountains. With the rover, you could do that. It took us 40 to 50 minutes to drive down south; and I was the navigator. We had trained, so I was the navigator; and John was the driver of the rover.

We landed within a couple of hundred metres of where we thought we were going to land. So we, you know, basically recognized the major landing spots. And I remember as John started off, I said: 'Okay, John. Steer 120 degrees for 1.2 kilometres, and then turn left to 090 degrees and go another 2 kilometres' or whatever it was. And so, that's the way we navigated. The Lunar Rover had a little directional gyro. There was no magnetic field on the Moon, so a magnetic compass wouldn't work. So we had a little gyroscope that was mounted in the instrument panel of the rover. We had a little odometer on the wheel that counted out in kilometres, and so that was our distance. And so, that's how we navigated up on the lunar surface. We'd start out one direction and we'd make a big loop and come back to the Lunar Module six to seven hours later. That was the plan. And, you never really worried about getting lost up there because everywhere you drove, you left your tracks. And so, if you really were unsure of your position, it was easy just to turn around and follow your tracks back.

We kept jogging and jogging, and the rock kept getting bigger and bigger and bigger. And we were going slightly downhill, that we didn't sense at first, and so we get down to this thing and we called it 'House Rock'. You know, it must've been 90 feet across and 45 feet tall. It was humongous. And we walked around to the front side or the east side, which was in the sunlight, and, you know, it was towering over us and John and I hit with a hammer, and a chunk came off, and we were able to collect a piece of House Rock. But – then we had to hike back. It was uphill, and it was a struggle getting back up.

Before he left, Charlie Duke left on the surface a picture of his family. A message on the back reads: 'This is the family of Astronaut Duke from Planet Earth. Landed on the Moon, April 1972.' Underneath were the signatures of his wife and children.

Apollo 17 – The Last Moon Landing

Probably the most challenging landing on the moon was the last, *Apollo 17*. On 6 December 1972 Gene Cernan, Harrison Schmitt and Ron Evans were in the Command Module on top of a fuelled *Saturn V* for the first night launch in US spaceflight – a time dictated by the angle of the Sun at the landing site in the Taurus-Littrow Highlands – the shadows cast by the craters must be low enough to show sufficient relief to allow a safe landing.

Apollo 17 astronaut Harrison Schmitt is dwarfed by a boulder during the third EVA at the Taurus-Littrow landing site.

Gene Cernan was the commander. He was making his third trip into space and his second on an Apollo mission. This time he was to land on what was to be the final Moon landing, along with geologist Harrison Schmitt. Originally Joe Engle was due to accompany Cernan, but when it was realized that *Apollo 17* would be the last such mission NASA decided to replace him with a trained geologist who had been preparing for the now-cancelled *Apollo 18*.

Gene Cernan: *Our valley where we were to land in was surrounded by mountains on three sides that are higher than the Grand Canyon is deep, to give you some idea. So at 7000 feet we were down among them. I mean the mountains rose above us on both sides. The valley was only 20 miles long and about five miles wide. We had good photography. So I practised this 100, 500 – I don't know how many times. So what I was looking at I'd seen before basically, because of the simulation and the pictures. So I knew we were in the right spot. At 7000 feet as the craters and rocks and the boulders and so forth began to appear I could begin to pick up my landing site. We had a particular target point, but it was only as good as we expected it to be. But when I got closer and I could see, then I could what we called redesignate where we were going to land. As I say, all the way down the engine is firing, you're in a suit, it's noisy, it's vibrating, people are talking to you from both ends, needles are going left and right. You know you don't have much fuel. So you got to get down quickly. But you can't get down too quick, you got to have your rate of descent under control.*

You get down to 200 feet and you're going to land or crash because if something happens to the descent engine at that point in time you can't react quick enough to stage the two vehicles, fire the ascent engine, and get out of there. So when you're down below 200 feet you're going to land. I mean I wasn't going to go all that way a second time and not land. Fortunately everything worked well for us. The landing radar, all the equipment, everything worked fine. You're coming down pretty fast through 200 feet. I don't remember exactly, but somewhere around 30, 35 feet per second, which is pretty fast. You've got to slow down from that point on so that you touch down at one or two feet per second. You get to about 80 feet, and you start blowing dust all over the place. By that time you now know where you're going to land, the dust keeps you from really seeing much of anything, because it just scatters horizontally in all directions. You effectively take what you got. It could have been two seconds, ten seconds, a minute or two, I don't know. But after we all got our breath and realized hey we are there, that's when I told Houston: 'Houston, the Challenger has landed.' There have been people who want to believe in the fantasy or the conspiracy, whatever, that it was all done in Hollywood, we never really walked on the Moon. Well, if they want to have missed one of the greatest adventures in the history of mankind, that's their choice. But once my footsteps were on the surface of the Moon, nobody, but nobody, could ever take, and to this day can take those footsteps away from me. Like my daughter's initials I put into the Moon during that three days we were there. Someone said: 'How long will they be there?' I said: 'Forever, however long forever is.' I'm not sure we, any of us, understand that.

As Cernan and Schmitt ranged over the Moon, back in Houston Cernan's daughter Tracy was interviewed on 'The Today Show'. She was asked if her father was bringing her back something special. Eventually she confided: 'He's going to bring me back a Moonbeam.' Soon it was time to leave the Moon for the final time. With a waning gibbous Earth in the black sky above him, Cernan said goodbye.

Cernan: When I climbed up the ladder for that last step, and I looked down, and there was my final footsteps on the surface, and I knew I wasn't coming back this way again, somebody would – and somebody will – but I knew I was not going to come back this way. I looked over my shoulder because the Earth was on top of the mountains in the southwestern sky. Never moved for the whole three days we were there. People kept saying: 'What are you going to say, what are going to be the last words on the Moon?' I never even thought about them until I was crawling up, basically crawling up the ladder. As I take man's last step from the surface, back home for some time to come – but we believe not too long in the future – I'd like to just say what I believe history will record – that America's challenge of today has forged man's destiny of tomorrow. And as we leave the moon at Taurus-Littrow, we leave as we came and, God willing, as we shall return with peace and hope for all mankind.

On the leg of the descent stage of the Lunar Module's leg is a plaque that reads: 'Here Man completed his first explorations of the Moon. December 1972 AD. May the spirit of peace in which he came be reflected in the lives of all mankind.' Nearby, next to the abandoned lunar rover, drawn in the lunar soil, are the initials, TDC, standing for Tracy Dawn Cernan, who was waiting to welcome her father home. *Apollo 17* splashed down on 19 December 1972. Since then, no one has left low-Earth orbit.

A Proud Legacy

Twenty-nine astronauts took part in the Apollo Moon programme. Between December 1968 and December 1972, 24 of them went to the Moon (three of them twice) and 12 walked upon it. At the time they walked, Charlie Duke was the youngest at 36, while Alan Shepard was the oldest at 47. Of the 12 moonwalkers, nine are still living. They are elderly now. At the time of writing Buzz Aldrin is the eldest at 78. According to NASA's latest plans the next manned flights to the Moon will begin not before June 2019 with the landing of the *Altair* lunar lander. By that time the youngest of the living moonwalkers will be 83 years old and the last man to walk there will be 85. I hope they enjoy the return.

'There's something on the telemetry that doesn't look quite right'

THE FIRST AMERICAN SPACE STATION

SKYLAB AND SOYUZ 11–13
1973–1974

The Americans' initial plans for an orbiting space station, conceived back in the 1950s, had been superseded by the rush to get to the Moon. But with lunar ambitions achieved for the time being, thoughts again turned to a US space station, and *Skylab* was launched in 1973.

The *Salyut 3* space station was sent into space on 25 June 1974 – another Almaz-type military station but one that was a considerable improvement on its predecessors, having superior living and working quarters. It had a floor and a ceiling painted different colours to provide the crew with some degree of orientation. Shortly afterwards, the crew of *Soyuz 14* arrived at *Salyut 3*, their 15-day mission going relatively smoothly. *Soyuz 15* was also intended to visit *Salyut 3* (carrying Colonel Lev Demin, at 48 the world's first space grandfather), but it had to return after two days due to docking problems. Despite this, Soviet planners were encouraged by their latest space station. Nevertheless, it was not visited again and re-entered the Earth's atmosphere in January 1975.

Skylab

The first space station launched by the United States was *Skylab*, which was visited by three crews between 1973 and 1979. Although Wernher von Braun had dreams of a space station in the 1950s, they did not get far before the race to the Moon took precedence. In 1963 the US air force started development of its Manned Orbital Laboratory (MOL), which was basically a small, two-man space station equipped with a large telescope for spying on the Earth's surface. Initially the MOL was to be launched as part of a *Titan II* rocket with a modified Gemini capsule on top. It was soon realized, however, that for reconnaissance work there was no need to have the space station manned, since automated systems would work just as well. It was in response to this US initiative that the USSR started its *Almaz* military space stations, some of which flew in space in the 1970s.

An overhead view of the *Skylab Orbital Workshop* in Earth orbit, as photographed from the *Skylab 4* Command and Services Modules (CSM).

At the time, NASA was worried about the possibility of a military space station and started looking at ideas of its own. It seemed logical to use the powerful Saturn booster being developed for moonflights and adapt Apollo hardware. Eventually it was decided to use an unfuelled third stage of a *Saturn V* rocket, which was fitted out with living quarters and a docking port. It was to be placed into orbit using a two-stage version of the *Saturn V*, left over from the cancelled Apollo missions.

It was launched on 14 May 1973 and immediately ran into severe problems. Among those watching its lift-off was Owen Garriott, who was to later live onboard it for 59 days.

Garriott: *The Saturn V launch, the big earth-shaker, went extremely well. We saw the take-off, saw it disappear, we were very pleased from the VIP stands. So we all came back to the motel, and I remember we changed our clothes into flight suits, because we were going to go out to Patrick Air Force Base to fly back to Houston in our T-38s. As we were coming by, walking out to get into our car, I noticed a gentleman on the second story up there, whose name was Rocco Petrone. Rocco had just been recently appointed the Director of the Marshall Space Flight Center and so we went up and said: 'Hey, that's great Rocco. Sure looks good.' 'Owen, don't get too excited. There's something on the telemetry that doesn't look quite right. It looks as if they're not getting the electrical power and the attitude is wrong, so we don't know just what's happening yet, but we have a problem.' So that's how I found out about the difficulty when the launch occurred and that Skylab's micrometeoroid shield had been torn away from the third stage of the* Saturn V.

Joe Kerwin had planned to lift-off in a *Saturn 1B* the following day with fellow astronauts Pete Conrad and Paul Weitz, to become the first crew to inhabit *Skylab*. Until, that is, he heard about the problems.

Kerwin: *Something went awry during launch with the heat shield. It seems the designers of the heat shield didn't talk to the aerodynamicists and they didn't properly protect the heat shield from the supersonic windstream. When the vehicle went supersonic, we got some windstream under the leading edge of the heat shield, and it just ripped right off the spacecraft. When it ripped around to one of the solar panels, it carried it off with it at the shoulder, so we lost one panel. When it got to the other panel, thank the Lord, it ripped around it instead and preserved that panel, but riveted it almost completely shut. A piece of aluminum with a rivet at one end literally wrapped it to this solar panel cover and held it down. So they got into orbit, commanded the solar panel covers to open. They got no response from one and they got a little trickle of power from the solar panel on the other side. The thing had opened, as it turned on, about a foot and then jammed. Then the temperatures inside began to climb.*

Meanwhile, it became apparent that afternoon that we weren't going anywhere the next day. This was not ready for a crew and might never be. So our first duty was to our families, who were having the pre-flight cocktail party at Patrick Air Force Base. We called them up and said: 'You can keep having your party, but we're not launching tomorrow. You can go on home.' In fact, the next morning, we got in our T-38s and flew back to Houston and joined the team that was going to try and figure out this problem and what to do about it.

On launch day, now May 25th, we showed up at the launch pad and there was practically nobody there. This was the least well-attended Apollo launch in history, because everybody had to go home and put the kids back in school, you know. So it was a very peaceful morning. We arrived at the Command Module and looked inside, and it was a sea of brown rope under the seats, and under the brown ropes were all these different umbrellas and parasols and sails and rope, and also the equipment that we had selected to try and free up the solar panel, which was a pretty eclectic collection of aluminum poles that could be connected together, and a Southwestern Bell Telephone Company

Owen Garriott, *Skylab* 3's science pilot, performing an EVA on the Apollo Telescope Mount (ATM) of the *Skylab Space Station*.

tree-lopper, with brown ropes to open and close the jaws, and all that stuff. They handed us the checklist and said: 'This is how to operate that stuff.' Some of it we'd seen, some of it we hadn't.

We then soft-docked with Skylab. I won't describe that in detail, but it's a partial docking. Sat there for an hour or so discussing with the ground – we had lunch at that point – what we were going to do. We decided that it was worth trying an EVA to open the side hatch of the Command Module and use one of the tools we had, which was sort of a shepherd's crook on a five-foot pole, to try and pry that solar panel up. We didn't know how much force it was stuck down with. So we tried it. We got our helmets and gloves back on, checked the suits out, and opened the side hatch. Paul Weitz had the shepherd's crook, I had Paul by the legs, and Pete, of course, was flying the spacecraft.

He'd fly it up within a couple of feet of the solar panel, and Paul would put his shepherd's crook out there and get it hooked under the free edge. He'd pull back, and the two spacecraft would come together, and there would be jet firings everywhere. Conrad would say: 'Oh my God.' We never collided or anything, but it was fairly sporting. Tried that two or three times, and it was obvious that it was stuck too hard and this was not going to work, unfortunately. So we closed the hatch, re-pressurized. Pete got himself a little ear block, I think, on that occasion, but he didn't say anything about it until the next day.

We said: 'Okay, that's not going to work. Let's go back and dock and just proceed with

1973

14 May US launches 85-ton *Skylab*, its first manned space station, with first crew Kerwin, Conrad and Weitz

September Russians restart their Soyuz programme, with the launch of *Soyuz 12*

18 December Russians launch *Soyuz 13*

1974

29 March US space probe *Mariner 10* completes a flyby of planet Mercury

26 December Russians launch their space station *Salyut 4*

activation, and we'll go to Plan B.' That's where we went back to dock and the docking wenches didn't work. Oh my God. They had worked the first time, for the soft docking. This time they didn't work. So here we are up there, now we'd been up for, I don't know, 18 hours or something already, and it's getting late. He tries it soft and he tries it hard. We tried all the backup means in the book. We finally got to the last backup, which required another EVA. It required that we de-pressurize the spacecraft again, because we now have to go up into the tunnel hatch, remove that hatch, get up into the docking probe itself, and cut a couple of wires and put the hatch back on. What it does is, this bypasses a relay that requires the capture latches to be mated before the main latches can activate. So we've solved that, and now he has to dock once more and keep, with the hand controller, keep the RCS [reaction control system] jets forcing the probe to collapse against the docking ring, and hopefully, when it collapses fully, the main latches will now latch on their own. So we go in, he's got it knocked, he's right in the middle. He's one, two, three, four – they said to give it ten seconds – five, six, seven, rat-rat-rat-rat-rat. There's machine gun noise, which is all 12 of those docking latches. We said: 'Thank God. We don't have to go home tomorrow.'

So the first two weeks, we were still in the dark. We still hadn't solved all the problems, so there was still tension. We were doing the exercise thing. We had a half-day off around day eight or nine, and we gave the ground a little television show. Weitz had a taped copy of 'Thus Spake Zarathustra'. It was the theme for 2001, Strauss. We played that real loud, and we set up the camera, and we came down into the workshop for the docking the day after and we did our acrobatics thing. See, we had learned how to do it in the first week, and the ground thought that was neat. So it relaxed everybody.

There were no handholds, no footholds, no visual aids, no lights, because there was never any planned maintenance on Skylab. Too dangerous. There was, fortunately, planned EVA. It was to retrieve film and exchange film in the ATM. So we had the suits, we had the umbilicals, and we brought up some tools that we thought we'd need. They planned an EVA that had us erect a twenty-five-foot pole, put the cable cutter on the end of it, and the jaws, which are about three inches long, had to close around that aluminum scrap that we'd seen, and bite halfway into it but not all the way. That was step one. Now we had a handrail. Pete could go along the handrail while I stabilized the near end of it, with another rope attached to his sleeve. When he got as far out as he could, taking care to avoid sharp edges, please, he would hook that rope into the solar panel cover as far down as possible, so as to give it some leverage from the hinge. Because what we had *to do was not only cut the scrap, but then break up that hinge which had frozen, and start the thing up. He was to go down, put the rope on, then I would tie the other end of the rope up to a stanchion, as close to the surface as possible, and then the two of us would get under the rope and stand up and hope for the best. That was the Rusty Schweickart solution. We said: 'Well, okay, Rusty, we'll give it a go.' And out we went with all the equipment. I even had a dental saw from the medical kit taped to the chest of my suit, just in case, if all else failed, we thought maybe we could go down there with the dental saw and try to get that thing off. Didn't have to use it.*

Gerry Carr visited *Skylab* for 84 days between 16 Nov 1973 and 8 Feb 1974:

The three main tasks, of course, of Skylab, were to study the human body, to study the Sun, and study the Earth. So we spent a lot of time on those experiments. The solar physics work was extremely difficult, because the Apollo Telescope Mount (ATM) control panel controlled a great number of different experiments. The Apollo Telescope Mount itself, the telescope, there were five or seven experiments in that big mount, and I always thought of it being like a big Gatling gun or a gun turret, because it turned. What you did is you would turn the drum inside there and you'd position one of the experiments to take solar data, and then when you finished with that, you would position another one and take data.

Russia Restarts the Soyuz Programme

It had been more than two years since a single Soviet cosmonaut had been in space following *Soyuz 11*. The newly designed *Soyuz 12*, launched in September 1973, was flown by two rookie cosmonauts, Commander Lieutenant Colonel Vasili Lazarev, 45, and Engineer Oleg Makarov, 40. Things were evidently normal during the first day of flight. Few scientific experiments were included in the programme. On the second day, however, serious defects developed in the life-support system, followed by a failure in the ship's attitude control system. The flight lasted barely two days.

Soyuz 13 was launched on 18 December 1973 and was the second test flight of the new Soyuz design. Both cosmonauts, Pyotr Klimuk and Valentin Lebedev, were rookies. Klimuk had trained for many years in the L1 and L3 lunar programmes before his assignment to the current mission. As the two cosmonauts entered orbit, it marked not only the 50th manned spaceflight but also the first time in the history of spaceflight that men from both the United States and the Soviet Union were in space at the same time. (NASA astronauts were in the middle of their marathon *Skylab 4* mission.)

Salyut 4, launched in December 1974, was a complete success, being visited by three Soyuz crews, including one that stayed for 63 days. Unmanned Soyuz capsules remained docked with it for prolonged periods, proving the Soyuz system's versatility and durability. The age of the Soviet space station had truly begun.

'We actually came to have a very close relationship with the Soviet crew'

EAST MEETS WEST IN ORBIT

THE APOLLO–SOYUZ TEST PROJECT
1975

The year 1975 saw the first international co-operative venture in space when American astronauts, taking part in the final Apollo flight, docked in space with Soviet cosmonauts in a Soyuz craft. Although there was a scientific dimension to the mission that would prove useful in the future Shuttle-Mir programme, the Apollo-Soyuz Test Project was largely a symbolic gesture of *détente* between the nations.

'Man this was worth waiting 16 years for,' said Deke Slayton, 51, as he blasted off with Tom Stafford and Vance Brand in the *Saturn* for the first international space mission. Slayton, one of the original Mercury 7, was making his first spaceflight, having been grounded for a heart murmur that was years afterwards regarded as trivial.

Although the international mission, a docking between an Apollo and a Soyuz capsule, was mainly about goodwill, for the American crew it was still welcome, since it was clear that it would be many years before there would be another flight opportunity with the Space Shuttle which was under development.

Speaking the Language

Vance Brand, 44, was also on his first mission and was the Command Module pilot. He recalled:

> At that time I'd just gotten off a backup assignment with Apollo 15, and I was a backup crewman on through the Skylab missions and very up to date on the Command Module and Service Module, things that would be used on the Apollo-Soyuz mission. I decided that I was interested enough that I attended some of the banquets that the Soviet cosmonauts were attending here. On my own, I went off, took Russian lessons. I paid for my own Russian lessons on Saturdays and got into their language and their culture a little bit.

Cosmonaut Aleksei Leonov (left) and astronaut Deke Slayton
(right) greet each other in space.

*When we got with the Soviets, they had their security monitors, and you could see that it was
a less trusting, more closed society, but on the other hand, as human beings [they] that opened up
more and more, I thought, in our relationships. We actually came to have a very close relationship
with the Soviet crew.*

A Handshake in Space

In orbit, the two craft docked and, after their cabin atmospheres were equalised (the Apollo used low-
pressure oxygen whereas the Soyuz used air at one atmosphere of pressure), the time came for the
handshake in space and the exchange of gifts.

Brand: *On the first transfer, Deke and Tom went into the Soyuz, and I stayed back in the
Command Module, sort of minding the store, so to speak, holding the attitude for the stack of
vehicles which consisted of Soyuz and Apollo and Docking Module. Tom and Deke went in, and,
of course, there was a big greeting – and they went into the Soyuz, they had something to eat,
signed some documents, and more or less made an international relations thing out of the first*

1975

15 July US astronauts Deke Slayton, Tom Stafford and Vance Brand blast off in an Apollo spacecraft, following Soviet cosmonauts Valeri Kubasov and Aleksei Leonov being launched in a Soyuz spacecraft

17 July The Apollo and Soyuz craft dock in space and the crews transfer between them for a time – the first such linkup

19 July The linked US and Soviet spacecraft separate

24 July The Apollo spacecraft splashes down safely in the Pacific Ocean

visit. Then later we had other transfers back and forth. I went over on one, and I was in the Soyuz for four and a half hours. Valeri Kubasov and I were together in Soyuz. Aleksei Leonov, on the other hand, was visiting Tom and Deke in our spacecraft. We couldn't freely go back and forth because of the airlock in between.

After the final undocking, they went to a lower orbit and speeded up and went ahead of us. We had tried to play a little joke on them at that point. Before the mission, on the ground, I had made a tape at home, and my daughter Stephanie and a friend helped me make it. We turned on shower water. They weren't in a shower, but it sounded like it. Both girls were about 18 or 19 years old, and so they made a lot of noise, which made it sound like somebody was in a shower just having a ball, a lot of giggling and stuff. So after the Russians were ahead of us, oh, 3 or 400 miles, and we had watched them go out ahead of us and probably tracked them a little bit, why, we played this tape over the VHF communications, which both spacecraft had, and we said: 'Hey, we're having a ball here,' and then we played this noise … And there were all these female voices and stuff. So I'm not sure that they heard that tape, actually, because after the mission Alexei was asked and he didn't act like he knew about it, but we tried, anyway, to play a little joke on them.

Forgetting to Throw the Switches

There were problems during the flight. Stafford forgot to turn off a switch and he flooded a cooling system. This meant that instead of staying constant, the temperature would swing erratically in the capsule. There was also the remote possibility that if the ice had expanded enough, it would have punched a hole in the cabin and they would have died. Whilst performing the docking, Deke Slayton hit a jet control handle and caused the command search module to yaw, and almost broke the docking interface.

It was also reported that Vance Brand forgot to throw two switches on the Earth landing system. The Apollo crew always wanted to operate these manually so that the parachutes did not deploy in orbit. The switches had to be thrown at 12,200 metres (40,000 feet). Throwing the switches shuts off the hydrazine thrusters that are firing to keep the Command Module stable. Because they remain firing, the pressure equalization that automatically sucks in air when the pressure outside gets greater than the pressure inside, also sucked in poisonous and highly corrosive nitrogen tetroxide.

Brand: *Nitrogen tetroxide is really a bad chemical to breathe because when it sees the moisture in your lungs, it turns into an acid. So we realized that we'd been gassed. I was right next to the vent, so I passed out momentarily after we got on the water, and Tom had us all put on oxygen masks. Then I came to, and we knew we needed to get the hatch open, to get fresh air in the cabin, but we weren't real quick to do that because we wanted to make sure the docking collar was around the spacecraft. We didn't want any water in the cabin. Eventually we got the hatch open and fresh air. We told the docs we thought we'd had some gas, so they checked us out. Sure enough, they could see it on our lungs, and so we were in the hospital on sort of a lung treatment protocol that was very good. It actually eventually reduced the irritation in our lungs, and I guess within two weeks I was jogging, and I haven't had any effects since then. So it was a nominal entry, except for that.*

Afterwards, the flight controllers had the two switches mounted on a piece of walnut with a little brass plaque. I'm told that Vance Brand still has them on his wall at home.

The mission was a success, and was the only time an American and Soviet spacecraft docked. When the Space Shuttle docked with the *Mir* space station for the first time in 1995, the USSR no longer existed.

Soviet Success

After all the failures and catastrophes of the Soviet lunar programme and the Soyuz accidents, the USSR only slowly exorcised the demons of its space programme. In 1977 *Salyut 6* would finally put the Soviet space programme on the slow track to success, hosting 16 crews, four of which set absolute endurance records for time in space, significantly exceeding the 84-day record set by NASA's *Skylab 4* crew during 1973–4. Space docking, and supply and refuelling, were developed to be routine. There were no fatalities, and in rescuing one of their space stations, *Salyut 7*, the Soviets pulled off one of the most remarkable feats in the history of spaceflight.

'Hey, there's some tiles missing back there'

THE MOST DANGEROUS MISSION OF ALL

THE SPACE SHUTTLE *COLUMBIA* (STS-1)
1981

When the Space Shuttle *Columbia* was launched in 1981, it ended a six-year period during which no American had gone into space. In that time, however, there had been 21 Soviet missions. The Soviets had, at last, reached their stride and were operating almost routinely, plying between the Earth and their series of space stations. However, the Americans could hardly have chosen a more dangerous mission to mark their return to space travel.

So far in the story of manned spaceflight we have seen many remarkable successes, as well as some disasters and near-disasters. Most accidents involved Soviet spacecraft and were caused by inadequate technology, poor manufacturing standards or pressure from politicians. But now we encounter American astronauts embarking on what many regard as the most dangerous spaceflight in history. For the first time, and unlike on all previous missions, they are using hitherto untried technology. There has never been a mission as risky as the first flight of the Space Shuttle. When Alan Shepard, John Glenn, Neil Armstrong and all the other American astronauts were launched on a *Redstone*, *Atlas*, *Titan* or *Saturn* rocket, they did so knowing that these rockets had been tested and approved before they climbed aboard. They also had escape systems, so that if anything went wrong they stood a good chance of survival. Things were different with the Space Shuttle. The first crew, veteran astronaut John Young and rookie Bob Crippen, would make spaceflight history by riding a rocket into space on its very first launch.

Untried Technologies

Bob Crippen, 43, had been selected as part of the second group of astronauts for the USAF Manned Orbiting Laboratory programme, but when that project was cancelled he moved to support the *Skylab* missions. He wasn't sure why he got picked for spaceflight's most dangerous mission.

> Crippen: *Beats the heck out of me. I had anticipated that I would get to fly on one of the Shuttle flights early on, because there weren't that many of us in the astronaut office during that period of*

time. I was working like everybody else was working in the office, and there were lots of qualified people. But one day we had the Space Shuttle Enterprise *(an engineering test vehicle not designed to go into space) coming through on the back of the Boeing 747. It landed out at Ellington Field, Houston, Texas. I happened to go out there with George Abbey, who at that time was the Director of Flight Crew Operations. As we were strolling around the vehicle, looking at the* Enterprise *up there on the 747, George said something to the effect of: 'Crip, would you like to fly the first one?'*

The Space Shuttle *Columbia* (STS-1), on the launch pad. The initials STS stand for Space Transportation System.

'Crip, would you like to fly the first one?'

GEORGE ABBEY

Almost everything about the mission was new and risky. It would have been relatively straightforward to make the first few flights of the Space Shuttle unmanned – even today, if all goes well, the only thing that the commander of the Shuttle must do during lift-off and ascent into orbit is throw one switch; the rest is automatic. But neither NASA nor the astronauts wanted that. Perhaps it was because it was so long since an American had been in space, or perhaps it was because so much could go wrong on the flight, that an astronaut was needed. Nevertheless, there were some who wondered if a six-year gap had led to complacency. It is true to say that many fellow astronauts feared for the lives of the crew of STS-1.

The only thing that had been tested on the Space Shuttle was the latter part of the landing. Four flights off the back of a Boeing 747 carrier aircraft from 7620 metres (25,000 ft) had been successful. The rockets of the Space Shuttle's main engines (SSMEs) – three of them on the rear of the Orbiter (the part that returned to Earth), as well as the two on the side-mounted, strap-on solid-fuel boosters (SRBs) – had only been tested on the ground, never in action with a crew on board. Furthermore, the SRBs had only ever been tested lying on their side; they had never been fired standing up. The massive external tank that holds liquid hydrogen and oxygen fuel had never been through the stresses of a launch. The Orbiter's heat shield, a 24,000 mosaic of silica tiles glued to its underbelly, had never experienced the 17,000 miles per hour (27,360 km per hour), 1600 °C re-entry. Many astronauts called it a fragile, glass spacecraft.

What's more, the Space Shuttle relied on its computers and hundreds of thousands of lines of computer code. Could all that be tested properly? The Space Shuttle was something new: a partly reusable spacecraft system, validated by computer.

Once the reusable twin SRBs fired, there was no way to throttle them back. In the words of one astronaut, it was then not a question of if you go, just of which direction. The SRBs, along with the SSMEs burning four million pounds of propellant in just over eight minutes would, according to the ground tests and computer models, propel a quarter-million-pound Orbiter into space to an attitude of 200 miles (322 km) and at a speed of 5 miles (8 km) a second. On its return it would re-enter the atmosphere half a world away from the final landing point and, without using power, fly through the ever-increasing air, decelerate due to friction, perform elongated S-bends, and then line itself up at the right heading, altitude and speed with a 4572-metre (15,000-ft) runway for a single-attempt landing.

No Escape Possible

But what if something went wrong? What would the crew do? The options were limited – very limited – and although no NASA PR official would admit it, then or now, the astronauts knew the score. The Orbiter was outfitted with SR-71 Blackbird ejection seats for its initial two-man crew. Theoretically they could be used during the first two minutes of flight, and again at the end of the mission when the Orbiter was below Mach 3 and at 3048 metres (100,000 ft), about ten minutes before landing. But in reality few believed they would be of any use.

Bob Crippen later said:

Well people felt like we needed some way to get out if something went wrong; in truth, if you had to use them (the ejector seats) while the solids were firing, if you popped out and then went down through the fire trail that's behind the solids, I don't believe that you would have ever survived, or if you did, you wouldn't have a parachute, because it would have been burned up in the process. But by the time the solids had burned out, you were up to too high an altitude to use it. On entry, if you were coming in short of the runway because something had happened, either you didn't have enough energy or whatever, you could have ejected. However, the scenario that would put you there is pretty unrealistic.

If a SSME failed it would be possible to land in Spain or Africa after a trans-Atlantic abort, but a failure during the two minutes when the SRBs were firing was a different matter. There were procedures, of course – so-called mission profiles flown in simulators. They had checklists that some astronauts referred to as something to read while you are dying.

Attempting to Launch

The first attempt at launch was postponed due to a computer hitch.

Crippen: *You know, the vehicle is so complicated, I fully anticipated that we would go through many, many countdowns before we ever got off. When it came down to this particular computer problem, though, I was really surprised, because that was the area I was supposed to know, and I had never seen this happen; never heard of it happening. It was where the backup computer couldn't hear what the primary computers were saying. There were four primary computers and one backup computer, and we considered the backup absolutely essential to have, but they weren't communicating properly. I know John and I spent an extra three hours out on the pad strapped in on our backs, for a total of six hours strapped in, before we finally gave up. In fact, that six hours is still used as the max to put people through, because it does get pretty uncomfortable strapped in on your back for that long a period of time. But we climbed out, and I said: 'Well, this is liable to take months to get corrected,' because I didn't know what it was. I'd never seen it. It was so unusual, and the software so critical to us. But we had, again, a number of people that were working very diligently on it. In fact, they proved what the problem was, which was an initialization thing. We just happened to catch a minute window when we started up the backup computer that caused the problem to occur. So it was rapidly concluded that: 'Hey, if you go do it again, the odds are it's not going to happen.'*

1981

20 February Space Shuttle *Columbia* fires its three engines in a 20-second test, clearing the final major hurdle to its maiden launch

10 April Maiden launch of the Space Shuttle *Columbia* is scrubbed because of a computer malfunction

12 April Space Shuttle *Columbia* launched, carrying astronauts Robert L. Crippen and John W. Young

14 April *Columbia* lands at Edwards Air Force Base, California, completing the first flight of America's operational Space Shuttle

14 May Russian *Soyuz 40* mission to *Salyut 6* space station marks the last flight of Soyuz spacecraft

A Successful Mission

The next launch attempt was on 12 April.

> Crippen: *About one minute to go, I turned to John. I said: 'I think we might really do it,' and about that time, my heart rate started to go up. I think they said it was – because we were being recorded, and it was up to about 130. John's was down about 90. He said he was just too old for his to go any faster. And sure enough, the count came on down, and the main engines started. The solid rockets went off, and away we went. First you want to make sure that the solids would do their thing, that the main engines would run, and that the tank would come off properly, and that you could light off the orbital manoeuvring engines as planned; that the payload bay doors would function properly; that you could align the inertial measuring units; the star trackers would work; the environmental control system, the Freon loops, would all function. So John and I, we were pretty busy.*

The Troublesome Tiles

Once the Space Shuttle was successfully in orbit, many astronauts back on the ground now believed, perhaps with a high degree of confidence for the first time, that they and the Space Shuttle had a future. The first task for Young and Crippen was to open the payload bay doors.

> Crippen: *I opened up the first door, and at that time I saw, back on the orbital manoeuvring system's pods that hold those engines, that there were some squares back there where obviously the tiles were gone. They were dark instead of being white. So I went ahead and completed opening the doors, and when we got ground contact we told the ground: 'Hey, there's some tiles missing back there,' and we gave them some TV views of the tiles that were missing. Personally, that didn't cause me any great concern, because I knew that all the critical tiles were the ones primarily on the bottom. But, of course, the big question on the ground was, well, if some of those are missing, are there some on the bottom missing? So I know there was a lot of consternation going on on the ground about, hey, are the tiles really there. But there wasn't much that we could do about it if they were gone, so I personally didn't worry about it, and I don't think John worried about it.*
>
> *We did our de-orbit burn on the dark side of the Earth and started falling into the Earth's atmosphere. It was still dark when we started to pick up outside the window; it turned this pretty colour of pink. It wasn't a big fiery kind of a thing like they had – with the Apollo Command Modules and those kinds of things, they used the ablative heat shield. It was just a bunch of little angry ions out there that were proving that it was kind of warm outside, on the order of 3000 °F out the front window. But it was pretty. I've often likened it – it was kind of like you were flying down a neon tube, about that colour of pink that you might see in a neon tube.*
>
> *The autopilot was on. It was going through the S-turns. John was somewhat concerned on that first flight that when we got down deeper into the atmosphere, whether those S-turns were*

going to work right. He ended up taking over control at around Mach 7. I deployed the air data probes around Mach 5, and we started to pick up air data. We started to pick up TACANS (Tactical Air Control and Navigation System) to use to update our navigation system. And we could see the coast of California. We came in over the San Joaquin Valley, which I'd flown over many times, and I could see Tehachapi, which is the little pass between San Joaquin and the Mojave Desert. You could see Edwards, and you could look out and see Three Sisters, which are three dry lakebeds out there. It was just like I was coming home. I'd been there lots of times. I did remark over the radio: 'What a way to come to California.'

The Sole Flight of the Russian Shuttle

The first shuttle mission was judged a success, and preparations were in hand for the second flight of *Columbia* – the first time a spacecraft had ever been reused – in six months' time. The success had not been lost on the Soviets, who noted the intention to launch Space Shuttles from the Vandenberg Air Force Base in California from where they could go into polar orbit and overfly every square inch of Soviet territory, carrying weapons in their payload bay. The Soviets decided they needed a space shuttle as well. In fact, they decided to copy the US one. The Soviet space shuttle flew once, on an unpiloted mission in 1988, orbiting the Earth twice and making a perfect landing. The project was cancelled shortly afterwards.

In the meantime the Soviets had launched a seven-day mission to the *Salyut 6* space station by two cosmonauts, Leonid Popov and Dumitru Prunariu of the Romanian air force. It was to be the last standard flight of the Soyuz spacecraft, which was to be replaced by the more capable Soyuz-T series. In reality, however, the Soyuz spacecraft, then and now the cornerstone of their manned spaceflight effort, had been undergoing a continuous evolution since its introduction. The Soyuz-T variant had a revised rendezvous system, uprated solar panels and allowed its crew of three to all wear spacesuits, something which had not been possible before.

Columbia returned to space on 12 November 1981 commanded by Joe Engle with Richard Truly as pilot. Unknown to the media or the public, it was to shock many astronauts and engineers.

The Soviet space shuttle on display at the 38th
Paris International Air and Space
Show in 1989.

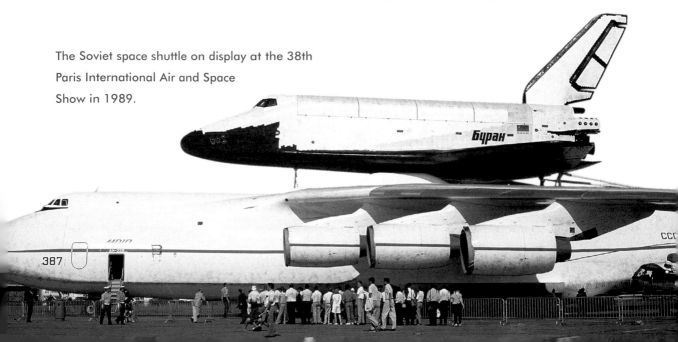

'It marks our entrance into a new era'

LAUNCH OF THE SPACE SHUTTLE PROGRAMME

COLUMBIA AND CHALLENGER (STS-2 – STS-9)
1982–1983

The maiden flight of the American Space Shuttle *Columbia* (STS-1) was followed by three research missions, each designed to evaluate the craft's technology and performance. *Columbia*'s first operational mission, STS-5, carried a four-man crew. But despite the euphoria and confident predictions that surrounded the programme, there were fears that the speed at which it was developing masked some important operational problems that had not been satisfactorily resolved.

Six air-worthy shuttles had been built. The first one, *Enterprise*, was intended for atmospheric tests and not meant to fly in space. Initially four spaceworthy shuttles would be built: *Columbia*, *Challenger*, *Discovery* and *Atlantis*. Later *Endeavour* would be added to the fleet.

STS-2: Damage to the O-rings

After a successful lift-off it was discovered that one of *Columbia*'s three fuel cells, responsible for generating electricity, had failed. Fuel cells were being used for the first time in a Space Shuttle, and with only two working, the decision was made to return to Earth after only two days. It was called a successful failure. But after the jettisoned solid-fuel reusable boosters (SRBs) were fished from the sea 150 miles (241 km) northeast of Cape Canaveral and sent back to the Morton-Thiokol plant for refurbishment, it was discovered that one of the eight rubber O-rings on the right-side booster showed heat damage. The SRBs consisted of sections, with an O-ring between the joins. However, the rings should not have suffered such damage. First, they were protected by the sheer size of the SRBs (3.7 metres/12 ft in diameter). Second, when firing, the SRBs burned from the inside out, meaning there was unburnt propellant between the rings and the heat. Third, the rings were protected by additional thermal insulation. Engineers had never seen this kind of damage before, either on STS-1 or in ground tests .

In retrospect, what followed set the course for a subsequent disaster. An O-ring was deliberately damaged and tested, showing that even in its damaged form it could withstand several times the pressure in a burning SRB. This caused the engineers to conclude that the situation was safe, despite not getting to the bottom of why the O-ring failed in the first place. Somehow, many engineers and managers came to believe that the O-ring damage seen in STS-2 was the maximum that would ever be sustained. No one ever told the astronauts about this; they just believed that SRBs were big, dumb and rugged. But soon they would all hear about damaged O-rings.

STS-3: Flaws in the Software

Columbia flew in space again the following March, 1982. There was no O-ring damage. Astronauts were beginning to understand the Orbiter and its sometimes pesky computer system. Gordon Fullerton, who was the pilot, remembers:

> *When we flew STS-3, we had a big book called* Program Notes, *which were known flaws in the software. There was one subsystem that when it was turned on, the displays said 'Off', because they'd gotten the polarity wrong and the logic, which they knew and they knew how to fix it, but we didn't fix it. We left it as it was and flew it that way, knowing that 'Off' meant 'On' for this subsystem. The crew had to train and keep all this in mind, because to fix it means you'd have to revalidate the whole software load again, and there wasn't time to do that. They had to call a halt and live with some real things you wouldn't live with if you'd bought a new car.*

**Space Shuttle *Challenger* atop its carrier aircraft –
a modified Boeing 747 – flying over Johnson
Space Center in April 1983.**

STS-4: Space Shuttle Goes Operational

1981

12 November Shuttle STS-2 is launched before landing back at Edwards Air Force Base, the first time a manned space vehicle has been reused

1982

22 March Shuttle STS-3 is launched before landing back at White Sands Space Harbor

27 June Shuttle STS-4 is launched, the final testing flight without a crew on board

11 November Shuttle STS-5 is launched with a four-man crew, to deploy the first commercial satellite in space

1983

18 June Shuttle STS-7 is launched, carrying the first American woman into space

28 November Shuttle STS-9 is launched, carrying a six-man crew and *Spacelab 1*

STS-4 was the first shuttle mission to be launched on time when it lifted off at the end of June 1982. On board, it carried many scientific payloads as well as a secret experiment for the US air force. Its two SRBs were lost when their parachutes failed to deploy, causing them to hit the water at high speed and sink.

When *Columbia* touched down at Edwards Air Force Base in California after the seven-day mission (for the first time on a concrete, as opposed to a saltpan, runway), President Ronald Reagan was there to greet it (his ranch was an hour's helicopter flight away). After Reagan had greeted the returning astronauts, the head of NASA, Jim Beggs, introduced the president to the awaiting crowd quoting from Reagan's inaugural speech: 'great nations dreaming great dreams'.

The president did not disappoint:

> *The fourth landing of the* Columbia *is the historical equivalent to the driving of the golden spike which completed the first transcontinental railroad. It marks our entrance into a new era. The test flights are over. The groundwork has been laid. And now we will move forward to capitalize on the tremendous potential offered by the ultimate frontier of space. Beginning with the next flight, the* Columbia *and her sister ships will be fully operational, ready to provide economical and routine access to space for scientific exploration, commercial ventures and for tasks related to the national security.*

After just five hours of flight, four lift-offs, eight SRBs used (including two SRBs lost) and 12 SSMEs firing, Reagan declared the Space Shuttle 'operational'. He also hinted that a commitment to a US space station was imminent. Sitting on top of a Boeing 747 carrier at the far end of the Edwards runway was the freshly built *Challenger*. Reagan continued:

> *If you'll all just look — well, I'm sure down in front maybe you can't see — but way out there on the end of the runway, the Space Shuttle* Challenger, *affixed atop a 747, is about to start on the first leg of a journey that will eventually put it into space in November. It's headed for Florida now, and I believe they're ready to take off.* Challenger, *you are free to take off now.*

Moments later one doomed Space Shuttle flew overhead, leaving the other doomed spacecraft behind.

STS-7: The First American Woman in Space

In June 1983 Space Shuttle *Challenger* carried the first American woman into space. Her name was Sally Ride. Rick Hauck was the mission's pilot. He recalled:

> *We'd have press conferences, and Sally would be the focus of 99 percent of the questions, but that was fine. I remember one press conference just before we flew. Someone from* Time *magazine or something said: 'Sally, do you think you'll cry when you're in orbit?' And of course, she kind of gave him this 'You gotta be kiddin' me' kind of look and said: 'Why doesn't anyone ever ask Rick those questions?'*

'Sally, do you think you'll cry when you're in orbit?'

QUESTION FROM *TIME* MAGAZINE REPORTER

STS-9: A Traumatic Landing

On 8 December 1983. *Columbia*, at the end of STS-9 (also known as STS-41A), was heading for touchdown at Edwards. On approach, one of its hydraulic pumps sprung a leak and highly inflammable hydrazine sprayed into the aft engine compartment. The fire spread to a second hydraulic system. Both quickly failed, followed by the third. On the flight deck Commander John Young noticed that the controls had frozen. Fighting to regain some control, he yanked the joystick from side to side, pressed his feet on the rudder pedals and applied the speed brake, but all to no avail. *Columbia* twisted as it descended, striking the runway on its side. Its wing buckled, tiles fragmented and splayed in all directions, the payload bay doors were torn and fire engulfed the craft's rear end.

Spacelab 1 had been the ninth Space Shuttle flight and *Columbia*'s sixth. John Young was making his sixth spaceflight. There had been a one-month launch delay because of a problem with a nozzle on one of the SRBs. *Spacelab 1*'s six-member crew was a record at the time. Brewster Shaw was the rookie pilot and Owen Garriott returned to space for the first time since *Skylab*. Ulf Merbold and Byron Lichtenberg were the first two non-astronauts to fly. Merbold was German, chosen by the European Space Agency that built the *Spacelab* – a multisubject laboratory fitted inside Columbia's cargo bay. They had been in space for ten days – the longest Space Shuttle mission to date. The crew worked 12-hour shifts in two teams, performing a variety of experiments spread over many disciplines, including life sciences, materials sciences and astronomy.

Even before the traumatic landing, *Columbia* had experienced other serious problems. Four hours before re-entry one of *Columbia*'s guidance computers crashed when a thruster was fired. Moments later a second computer (of five) crashed. John Young later said that his legs turned to jelly when he saw it. He delayed the landing, later saying in his matter-of-fact understatement: 'Had we then activated the backup flight software, loss of vehicle and crew would have resulted.' Young and Shaw worked the checklists and successfully rebooted the errant computers. Post-flight analysis showed a loose piece of solder on a CPU board was to blame. *Columbia*'s crew overcame these problems and also those encountered just before landing, and the Space Shuttle managed to touch down successfully. But John Young and his crew had been within seconds of death. The Space Shuttle was clearly not yet fully operational after all.

'You could see the Sun lighting the desert way up ahead'

FIRST COMMERCIAL SHUTTLE FLIGHTS

STS-41B, STS-41C, STS-41D AND STS-51A

1984

The year 1984 was significant in several respects for the Space Shuttle programme. The five missions which took place that year included the deployment – and recovery – of satellites, the first space flight with two women among the crew and the first spacewalk by a female US astronaut. Yet despite successes, the continuing failure of the infamous O-ring seals on the giant booster rockets were a chilling portent of a shocking catastrophe still to come.

The next Space Shuttle mission, STS-41B, took place in February 1984. It was *Challenger*'s fourth flight. It was Vance Brand's third spaceflight and his second Space Shuttle flight, having been part of the Apollo-Soyuz Test Mission as well as Shuttle mission STS-5 in November 1982. For him the Shuttle's re-entry was a special event. He recalls:

> In the early part of entry the vehicle had a large angle of attack as the Shuttle was pitched up roughly 40 degrees with respect to the velocity vector. There were very large windows, and you weren't looking backwards at a doughnut of fire as with Apollo. You were able to see the fire all around you, and you could look out the front.
>
> First the sky was black during entry, because you were on the dark side of the Earth, but as this ion sheet began to heat up, why, you saw a rust colour outside, then that rust colour turned a little yellowish. Eventually, around Mach 20 you could see white beams or shockwaves coming off the nose. If you had a mirror – and I did on one of my flights – you could look up through the top window, which was a little behind the crew's station, and see a pattern and the fire going over the top of the vehicle, vortices and that sort of thing. So it was really awesome.

Astronaut Bruce McCandless tests a mobile foot restraint during Space Shuttle mission STS-41B.

At Mach 18, or 18 times the speed of sound, I had a manual control task, which was to take over manual control from the autopilot and do a flight test manoeuvre. First I pushed down from 40 degrees to 35 degrees angle of attack, then up to 45 and then back. But when I did that, Joe Allen, who was sitting in the centre seat as flight engineer, was watching what was coming off the nose, the shockwaves, and he said a shockwave came from the nose and it came up and attached to the window right in front of us. That was a little worrisome, because he knew it was hot.

Eventually we were coming over the coast of the United States. I guess when we first landed at Edwards, it was dawn, but you could see the Sun lighting the desert way up ahead. When you get down to about Mach 3 to 2 you're getting into thick atmosphere and it's rumbling outside. You can hear it rumble, and you're decelerating such that it's pushing you into your seat straps. At Mach 1, you feel a lurch, and increase and decrease in deceleration and a decrease as you go through. Eventually you're over the field. On STS-5 we went through a very thin cloud deck. I was on instruments flying and circled down and landed at Edwards. We had an intentional max braking test and completely ruined the brakes. I had to stomp on them as hard as I could during the rollout, which points out that we had a lot of flight testing on that mission. Even though it was billed as the first commercial flight, I think we had roughly 50 flight test objectives.

We had a strange flight. We had a lot to do, the EVAs went just famously. It was the first time that people had flown with the Martin backpack Manned Manoeuvring Unit (MMU). It was supposed to be an early Buck Rogers flying belt, if you know what I mean, except it didn't have the person zooming real fast. It was a huge device on your back that was very well designed [and] redundant so that it was very safe, but it moved [you] along at about one to two or three miles per hour. It used cold nitrogen gas coming out in spurts to thrust you around.

> 'It was supposed to be an early Buck Rogers flying belt ...'
>
> VANCE BRAND

Satellites on the Wrong Path

Part of the mission was to deploy two satellites, *Westar* and *Palapa B-2*. Although the satellites were released, unfortunately they went into the wrong orbits.

Brand: *The deployments didn't work out so well. Here we were deploying two satellites that were similar to what were so successfully deployed on STS-5. One, the* Palapa, *was for Indonesia, and the other one was for Western Union. Anyway, it turned out that their booster rockets had similar failures. The deployments from the payload bay went flawlessly. Everybody checked that backwards and forwards. But what happened about a half hour after each of the satellites left us, was it started its burn, which was to send the satellite up to an orbit 23,000 miles out away from the Earth, called geosynchronous orbit. The engine burns, which were solid rocket burns, each started, and then after about 20 seconds unexpectedly stopped. We had the underside of our vehicle pointed in the direction of the burning satellites so that any speeding particles from the burn would just hit the underside and wouldn't do any harm. We were observing with a TV camera on the end of the Shuttle's arm. I'm not sure even today that it's well understood why those rockets burned out prematurely, but each left its satellite stranded in an inappropriate orbit. Of course, one of the satellites was later rescued by a follow-on EVA mission.*

STS-41C: Repairs in Orbit

The next Shuttle mission, STS 41-C in April 1984, was also dramatic. It was designed to capture and repair an ailing scientific satellite – *Solar Max*, built to study the Sun. It had malfunctioned, but since it was designed to be repaired in orbit *Challenger*, on its fifth flight, was sent to undertake the task. Because *Solar Max* was tumbling, astronaut George Nelson flew the MMU (manned manoeuvring unit) backpack out from *Challenger*'s cargo bay in an untethered spacewalk, carrying a capture device.

> Nelson: *Everything worked perfectly until I got to the satellite and flew up to dock with it, and then it didn't work. So I ended up making things worse, making the satellite tumble, and trying all kinds of stuff, actually just grabbing hold of the solar arrays. It was pretty exciting in retrospect, and the memories of the view from there are just amazing; the Shuttle against the Earth and jets firing and all this. What an extraordinary experience to be able to fly the MMU.*

Fortunately it proved possible to grab the satellite with the Shuttle's robot arm and bring it into the cargo bay where it was repaired and returned for several more years of productive scientific life.

STS-41D: Another O-ring Problem

In late August 1984 the 100th manned spaceflight took off with the first flight of *Discovery*. Carrying a crew of six, it successfully deployed three satellites. But with hindsight this successful mission will be remembered for things other than the satellites it deployed. During launch an O-ring in each SRB failed. Flames had penetrated their way to the casing joints of the SRBs and started to eat away at the rubber seal of the backup O-ring. The leak on the left-hand SRB was so bad that hot gas broke through the primary O-ring. The SRBs did not fail before they burnt out and were jettisoned, but had they burned for just a few more seconds the O-ring would have been breached and the crew would have perished. The crew were unaware of the dangerous situation, however, and only found out after the *Challenger* investigation.

Fear of Flying

Mike Mullane was a mission specialist on *Discovery*'s first flight, which was scrubbed several times before it launched. He recalls:

> *There is nothing that is more exhausting than being pulled out of that cockpit and knowing you have to do it tomorrow. It is the most emotionally draining experience I ever had in my life of*

actually flying on the Shuttle. I will admit that it is terrifying to launch. Once you get up there, it's relaxing, but launch, it's terrifying. And people assume that it gets easier. I tell people, no, it doesn't. I was terrified my first launch. I was terrified my second launch. I was terrified my third launch. And if I flew a hundred, I'd be terrified on a hundred. And as a result, you have this sense of death. You think about it a lot before you go fly. You prepare for death, basically. I know it's ridiculous to think you can predict your death. You could get in an auto accident driving out to get in the T-38, and that's your death, and here you are thinking it's going to be on a Shuttle. But I certainly prepared for death in ways, in a formal way. I served in Vietnam, and there was certainly a sense of you might not come back from that. And I said my goodbyes to my parents and to my wife and young kids when I did that, but this time it was different because it's such a discrete event. It's not like in combat where in some missions you go off and fly and never see any enemy anti-aircraft fire or anything. But this one you knew that it was going to be a very dangerous thing. And as a result, 24 hours before launch, you go to that beach house and you say goodbye to your family, to the wife, at least. That is incredibly emotional and draining ... it could be the last time she's ever going to see you, and you know it's the last time you might ever see her.

Salyut 7: Fully Fitted for Life on Board

In April 1982 *Salyut 7* was launched. It was originally the backup to *Salyut 6*. There were delays with the follow-on *Mir* space station project, so it was decided to launch it as *Salyut 7*. With *Salyut 7*, the Soviets demonstrated that they could live and work in space for long durations, coping with almost every eventuality. *Salyut 7* had a docking port at either end of the station and it carried three solar panels. It had electric stoves, a fridge and constant hot water. It was first visited, in May, by the *Soyuz T-5* crew. They stayed for 211 days. The following year the two-man crew of *Soyuz T-9* stayed for 150 days and in 1984 three cosmonauts lived on board for 237 days.

> *'I will admit that it is terrifying to launch.'*
>
> MIKE MULLANE

STS-51A: A Retrieval Mission

There followed the cancelled mission 41E, and the successful sixth fight of *Challenger* 41G, which lifted off on 5 October 1984. Shuttle mission 51-A was the second flight of *Discovery*, launched less than a month after the previous mission had deployed the *Earth Radiation Budget Satellite*. Its crew of five were a few months into their training when the satellites deployed during the February STS-41B Shuttle mission malfunctioned and went into useless orbits. The two satellites were valued at several hundred million dollars and, although they were insured, NASA agreed to retrieve them, since it wanted to demonstrate the capabilities of the Space Shuttle. The McDonnell Douglas Corporation, which had made the satellites' rocket motors, were also very interested in getting them back so that they could discover what had gone wrong. So what was due to be a straightforward flight, the deployment of two satellites, became the adventurous retrieval of two wayward satellites as well.

The first idea was to use the Shuttle's robot arm to grasp a small stud nicknamed the snubber. The satellites would then be drawn into the payload bay. Rick Hauck, making his second flight, was the commander.

Hauck: *I remember one morning Dale Gardner (the flight's mission specialist) came in and he said: 'I was up all night. I was thinking this is not going to work. It's too small, the snubber's too small, the rates are too high, and we got to do something else. There's got to be a better way.' And he came up with the idea of developing what was later called the stinger, and the stinger was mounted on the MMU and essentially was a probe that could be driven up inside the solid-rocket motor cone, the rocket already having been fired, so that wasn't an issue. The idea is, you push the probe up into the cone and then you release fingers that pop open and snag at the throat of the cone and then you spin a wheel down that screws it down and you got it. And it had the advantage [that] you didn't have to match any rates with the satellite. If you're going down the axis of rotation, you don't have to do anything other than be on the axis of rotation and then capture it. And then, of course, there's some rotation that's transmitted through the mechanism to the astronaut, but you just flip a switch on the MMU and nitrogen jets are fired and it's stabilized. So it was a brilliant idea.*

Safe Recovery

In space, *Discovery* manoeuvred alongside the first satellite to be captured, *Palapa B-2* (the orbits of both satellites had been lowered so that they could be reached by the Shuttle). *Palapa B-2* was recovered, with *Westar* recovered a day later. It was the last time a non-tethered spacewalk has been performed, and the last use of the MMU. It has been replaced by a smaller version called SAFER, which is used for emergency purposes only.

Hauck: *I saw a NASA press release or a statement from one of the people in the NASA Press Affairs Office saying: 'STS-51A. The crew's going to go up, they're going to launch a satellite, and they're going to bring back two satellites,' as if it was a piece of cake. And I was livid. I thought: 'Here we are, NASA is shooting themselves in the foot because we are implying that this is easy.' And I had the opportunity to see this gentleman the morning of the launch, and I said: 'You have set NASA up for a humongous failure by the nature of this press release.' And I said: 'In my view, if we get one of these satellites back, it'll be amazing, and if we get both of them back, it'll be a miracle.' And I said: 'You have not done NASA any favours.' Well, we got both of them back, thankfully, but we were close to getting neither of them back.*

1984

3 February Shuttle STS-41B is launched, the fourth flight of *Challenger*

6 April Shuttle STS-41C is launched, the fifth flight of *Challenger*. It recovers the ailing satellite *Solar Max*, designed to study the Sun

30 August Shuttle STS-41D is launched, the first flight of *Discovery*. Despite potentially catastrophic O-ring failures, the mission successfully deploys three satellites

5 October Shuttle STS-41G is launched. The mission is the first flight with two female crew members and includes the first spacewalk by an American woman astronaut

8 November Shuttle STS-51A is launched, the second flight of *Discovery*. The mission recovers two satellites deployed during mission STS-41, which went into incorrect orbits

'The most important thing for us was to dock'

RESCUE OF A CRIPPLED SPACE STATION

SALYUT 7, SOYUZ T-13 AND MIR

1985–1986

Evidence of growing Soviet confidence and technical abilities became apparent when a crew was sent to attempt to reactivate the Soviet space station *Salyut 7*, which had experienced a power shutdown. The following year, Soviet cosmonauts achieved another space 'first', by travelling from one orbiting space station to another several thousand miles away.

In 1985 the *Salyut 7* space station suffered major systems failures following the departure of the *Soyuz T-12* crew. The seventh expedition to *Salyut 7*, Vladimir Dzhanibekov, Svetlana Savitskaya – the second Soviet female in space – and Igor Volk, had stayed on board the space station for 12 days beginning in July 1984. Dzhanibekov and Savitskaya carried out a spacewalk, during which they tested equipment as well as cutting, welding and coating metal samples.

But now *Salyut 7* was without power and according to one Western commentator was 'dead in the water'. Even if the systems could be brought back on, it was by no means certain that a Soyuz craft could dock with it to enable a crew to get aboard. Nevertheless, it was decided to try – a clear indication of the growing confidence of the maturing Soviet space effort.

The Recovery Mission

Vladimir Dzhanibekov could not have imagined that he would be going back into space so soon, but he and the *Soyuz T-12* backup cosmonaut Viktor Savinykh were the obvious choices. They left Earth on 6 June 1985, and one day later they were closing in on the crippled space station.

> Dzhanibekov: *We saw the station directly after it had emerged into the light – it was blazing in the Sun's rays, which were just beginning to penetrate the atmosphere. A dot was not a dot, a speck was not a speck – they grew as we approached. The Moon was also within our field of vision. The Salyut's crimson colour gradually grew lighter and finally became white, the shade of ivory. The Salyut appeared to flare still more and at times was painful to look at through binoculars.*

The Russian *Salyut 7* space station in orbit, with ferry spacecraft *Soyuz T-14* docked at the bottom.

1985

6 June Soviet spacecraft *Soyuz T-13* launches. Its mission is to attempt to dock with, and return to operational status, the 'dead' space station *Salyut 7*

13 June Following earlier successful power restoration, *Salyut 7*'s attitude control system was reactivated, meaning that a *Progress* cargo craft could deliver new supplies

1986

19 February Soviet *Mir* space station is launched

5 May Cosmonauts begin a journey from *Mir* to *Salyut 7* – the first ever inter-space station flight

We made out the solar panels. At first they seemed to be correctly oriented towards the Sun – a ray of hope. But within a few minutes it was obvious that this was an optical illusion. They were facing in the opposite direction and were useless. It seemed clear that there would be big problems with the electricity supply. But the most important thing for us was to dock – the rest could come later.

We hard docked and checked the hermetic sealing. The equalization of the pressure on either side of the hatch and its opening gave us no problems. The only delay was in order to perform a gas analysis of the inside of the space station. It was possible that a short circuit had caused a fire in which case the scorched remains would have produced a dangerous atmosphere. So we sat there in the docking compartment patiently operating the level of the air sampler, watching to see if the indicators changed colour. Just in case, we had gas masks.

There was complete silence in the docking compartment and we were in semi-darkness. The rays of our torch picked out specks of dust hanging motionless in the air. This oppressive silence and the stillness of the air were the first signs of anything wrong in the station. Our most important compartment was the main compartment. There we found the same darkness, only now it was total. I removed my mask and sniffed the air. It seemed to be normal but the smell wasn't the usual one. It was a kind of stagnant factory smell. I took several photographs with flash and floated towards the table.

Awaiting us were some rusks and some salt tablets – the traditional 'bread and salt' welcome from the previous crew after their 237-day epic stay. We drew the blinds and let in daylight, everything seemed in order, clean and dry, no damage. The portholes were covered in frost. We were by now feeling the cold, the Salyut's internal temperature was below zero. The worst thing was that there was no electrical power and the back-up batteries were dead. The first thing we had to do was to link the batteries to the solar panels so that they could become charged. We drew up circuit diagrams and severed cables and improvised for insulation. A day after we had docked we rejoiced to see a rise in voltage as the batteries recharged.

Back to Life

Dzhanibekov: *Without ventilation, exhaled carbon dioxide accumulated around us. Imperceptibly we grew more and more tired and our heads began to ache. I taught Viktor to breathe out more energetically in order to expel the breath further away. He advised me to wave away the cloud with my hand. To combat the cold we wore fur coats, caps, gloves and boots.*

At first we worked only in daylight; it was uncomfortable to do so in the 45 minutes of orbital night. When it was dark we would float over to the Soyuz to get warm and to breathe properly. We realized that the space station was coming to life when we could turn the lights on. Then the instruments started working. It was a wonderful moment.

We still had a major problem – water. When we entered the Salyut we switched on the Rodnik water supply system but it was frozen and there was ice in the tanks. Then I noticed a strange plastic column that had grown onto one of the junctions of the water pipes. It was ice, water was escaping from the system and freezing in front of our eyes. When our week's supply of water would run out we were prepared to drink the coolant water from our spacesuits. We had collected drops of water from various pipes and hoses to add to our daily ration. We knew that our comrades on Earth were preparing an unmanned cargo craft to send to us containing plentiful supplies.

> ## 'Earth greeted me with sunshine and smiles.'
>
> VLADIMIR DZHANIBEKOV

The cooker did not work so we rigged up a system out of towels and foil and a powerful photographic lamp and used it to heat packets of food as well as coffee. The Salyut was slowly coming back to life.

The most difficult thing in the cosmonaut's trade is the capacity to stand up to, every day, minute by minute, the thousands of unforeseen circumstances. The arrival of the cargo craft, with the resultant unloading and loading, was more than just moving a few boxes. At one time, for instance, five spacesuits had accumulated on board and they took up a tremendous amount of space.

The spacewalk to charge up the third solar battery required several days training and the checking and adjustment of spacesuits. Soon we listened to the voices on the other side of the hatch signaling the arrival of our replacement crew – Volodya Vasyutin, Sasha Volkov and Georgi Grechko. I was to return, Viktor was to stay. Earth greeted me with sunshine and smiles.

A Year in Space

Thanks to the rescue of *Salyut 7* by Vladimir Dzhanibekov and Viktor Savinykh the space station was host to more long-duration crews, and on 6 May 1986 another space first was achieved.

The *Mir* space station had been launched on 19 February 1986. Leonid Kizim and Vladimir Solovyov docked with *Mir* on 15 March, on the *Soyuz T-15* mission. They stayed on *Mir* for 51 days, unloading two visiting *Progress* cargo craft. On 5 May they undocked and travelled in their spacecraft to *Salyut 7*, which was 2485 miles (4000 km) ahead of them in a lower orbit. It took them 29 hours, and it was the first ever inter-space station flight. The cosmonauts stayed on board *Salyut 7* for a further 51 days. Then it undocked and journeyed back to *Mir* for another 20 days, during which Kizim became the first person to have spent a year in space.

'Uh oh'

THE *CHALLENGER* DISASTER

STS-51L
1986

On 28 January 1986 the Space Shuttle *Challenger*, designated mission number STS-51L, blasted off from Cape Kennedy. For those watching, it appeared initially to be another exhilarating, yet almost routine, lift-off. Among the spacecraft's mission tasks was the deployment of the second in a series of tracking and data relay satellites. Also on board *Challenger* was a teacher, Christa McAuliffe, who was there as part of the Teacher in Space Project. But just over a minute into the Shuttle's flight, the unthinkable occurred.

In January 1986 Mike Mullane was in training for a secret military Shuttle mission to be launched from its alternative launch site at Vandenberg Air Force Base in California. He recalls:

> *I was with the rest of the STS-62, we were in training at Los Alamos Labs. One of our payloads was being developed at Los Alamos Labs, so we were up there at Los Alamos. We were in a facility that didn't have easy access to a TV. We knew they were launching, and we wanted to watch it, and somebody finally got a television or we finally got to a room and they were able to finagle a way to get the television to work, and we watched the launch, and they dropped it away within probably 30 seconds of the launch, and we then started to turn back to our training. Somebody said: 'Well, let's see if they're covering it further on one of the other channels,' and started flipping channels, and then flipped it to a channel and there was the explosion, and we knew right then that the crew was lost and that something terrible had happened.*

Danger Signs

In the two years before the *Challenger* accident there had been 15 missions, 12 in the previous year alone. The Shuttle was beginning to do what it was meant to do – or rather what the PR said it should do – that is, become a regular 'space truck', routinely plying between the ground and low-Earth orbit. In April 1985 the Shuttles *Discovery* and *Challenger* were launched just 17 days apart. There were spacewalks, some non-

Challenger at lift-off on mission STS-51L. The smoke seen at bottom right was the first visible sign that an SRB joint may have been breached.

tethered satellite deployments for the Department of Defense and the commercial satellite retrievals and in-orbit repairs. The robot arm worked impressively. There were also myriad experiments performed on the mid-deck and in the payload bay. But the high flight rate was straining the system in terms of engineering manpower and spare parts. There were specific warning signs for those who knew how to recognize their significance.

Between *Challenger*'s fourth flight – the STS-41B mission in February 1984 – and its final flight (its tenth) in January 1986, there were 15 successful Shuttle missions. In only three of those missions was there no visible damage to the SRB O-rings; in nine of the missions the burn-through was serious. One mission, 51C, was launched after a bitterly cold January night at Cape Kennedy. When the recovered SRBs were inspected, the O-rings were found to be severely damaged. The Shuttle fleet should have been grounded. Sooner or later its luck would run out.

Lucky Escapes

On the next flight, STS-51D, *Discovery* blew a tyre and the brakes seized on landing due to a strong crosswind. For the time being landings at Cape Kennedy were halted. Three months and three Shuttle missions later, *Challenger*'s central SSME failed during ascent. Fortunately it was straightforward to 'abort to orbit', reaching a safe but slightly lower orbit than intended.

The mission prior to *Challenger* – 61C (*Columbia*'s seventh voyage) – could also have been a disaster. A valve malfunction meant that one of the SSMEs could have exploded at main engine cut-off when the Shuttle reached orbit. The shower of debris from the explosion would have crippled the Shuttle, making it impossible to survive re-entry. Fortunately, the launch was scrubbed for another reason and the fault found. But even that was not the only potentially fatal flaw. A faulty drainback valve from the external tank that could have killed the crew was also discovered.

A Ride to Death

As well as the crew of specialist astronauts, *Challenger*'s final flight also had a passenger: the schoolteacher Sharon Christa McAuliffe. (She was not the first passenger aboard a Shuttle, however: STS-41C also had a passenger on board – Congressman William Nelson, who was ostensibly a 'payload specialist' but in reality was hitching a ride because of his influential political position. He was the second politician to fly on the Shuttle following Senator Jake Garn the previous April.)

Seventy-three seconds after lift-off *Challenger* exploded in a heartbeat just after Commander Dick Scobee gave the order: 'Go at throttle up.' Shortly afterwards pilot Mike Smith was heard to say: 'Uh oh.' An O-ring in the right-hand SRB had failed, and flame was visible through the SRB wall. The damaged SRB then pulled away from the external tank, causing it to fragment. There were more than a million pounds of propellant in it when it detonated. *Challenger* disintegrated at 14,630 metres (48,000 ft) but continued to climb to 18,288 metres (60,000 ft). In fact, there was no explosion in the true sense. Although the dramatic and massive release of oxidizser and fuel gave that impression, the liquids were actually vapourizing and burning. *Challenger* disintegrated, but it seems the crew were not killed instantly.

Six weeks later NASA found the crew cockpit. Many had hoped that it would never be discovered, but since it was in relatively shallow water, sooner or later it would have been encountered by a diver or snagged on a fishing net. The cockpit section of *Challenger* was well built and survived the accident intact. It can be seen in still-frames training cables. For a few seconds the G-forces on the crew would have been intense – 12 to 20 G – but then they would have swiftly fallen to about 4 G and then a few seconds after that down to zero G as the cabin would have been in free fall. There is evidence that some of the crew were conscious after break-up because three of the four personal egress air packs on the flight deck were turned on, and air consumed was consistent with breathing until impact with the sea. We do not know if the crew were conscious all the way to the impact. The air supply was unpressurized, so if the cabin had been breached they would, mercifully, have passed out within a few seconds. The cabin fell 19,812 metres (65,000 ft) in 285 seconds before striking the ocean. It was this impact, at 210 miles per hour (338 km per hour), which killed the crew.

On the evening of the disaster President Reagan cancelled the State of the Union speech for a week and gave a national address from the Oval Office. Written by speechwriter Peggy Noonan, it closed with a quote from the poem 'High Flight' by John Gillespie Magee Jr:

> *We will never forget them, nor the last time we saw them, as they prepared for their journey and waved goodbye and 'slipped the surly bonds of Earth' to 'touch the face of God'.*

The subsequent Rogers Commission that looked into the cause of the accident discovered that engineers' worries about the O-rings were not passed on to NASA managers at HQ in Washington, the astronauts in Houston or the launch director at Cape Canaveral on that fateful day. Afterwards one astronaut said that every new NASA administrator should be taken on the Shuttle. Then they would know what it was all about after they had been scared 'witless'. In the aftermath of the *Challenger* disaster there was a hiatus in Shuttle flights for 33 months.

'We will never forget them ...'

PRESIDENT RONALD REAGAN

'My God, that's a lot of damage'

STS-26 AND OTHER FLIGHTS
1988 AND BEYOND

After a gap of nearly three years, the Space Shuttle resumed flights in September 1988. Despite the general view that flights from now on were going to be safer than previously, even return-to-flight mission STS-26 was about to experience the type of life-threatening incident that was the dread of all astronauts.

Commander for the Shuttle's return to flight in September 1988 was Rick Hauck:

> I was absolutely thrilled that I was entrusted with that mission. I think every member of the Astronaut Office, probably without exception, wanted to be on that flight. I did tell my family: 'This will be the safest flight ever flown by NASA, STS-26.' What I did not say was: 'And that guarantees that I'm coming home,' because, of course, there's no guarantee. But I was comfortable that, within my view, everything had been done to prepare everything for that flight.
>
> A lot of people, when they hear you flew once before Challenger and twice after, assume you must've been scared more after Challenger, and I said no. I was terrified on my first launch, I was terrified on my second, and I was terrified on my third. Challenger did not change the fear factor at all. If anything, it was a very slight sense that it was safer on the post-Challenger missions than it was before because people were more focused. Disasters tend to do that, tend to focus folks. So I had this sense of maybe a little slight less apprehension about my second mission, although in lots of ways I was still terrified. Challenger didn't change a thing.
>
> Prior to 51L, we had never lost a crew after launch. They lost the Apollo 1 crew and, of course, astronauts had been killed in airplanes and car crashes and so on, but we'd never lost anyone in a spaceflight. So even though on STS-7 and STS-51A I knew this was dangerous, I kind of comforted myself with the thought: 'We've never lost anyone before, so we've got this wired; we know how to do this.' Well, that comfort could no longer be delivered by that thinking after 511.. I was convinced that everything had been done that could be done to prepare the machine and the crew and the software, but I knew that my good friends had died the last time a

STS-26 crew (from left to right): Mike Lounge,
Richard Covey, David Hilmers, Rick Hauck and George Nelson.

machine had launched. I do absolutely remember counting down after lift-off to solid rocket motor
burnout and two minutes and ten seconds after launch and the solid rockets are gone, and I
remember thinking: 'Well, glad they're out of the picture.' I didn't mention that – I forget. Launch
plus about 20 seconds, we did something called an SM alert, SM being a systems maintenance
minor alert, but it's something that was annunciated. It was a minor issue, but it sure got our
attention there for a period of time.

'I was terrified on my first launch, I was terrified on my second, and I was terrified on my third.'

RICK HAUCK

Heat Tiles Damaged on Launch

Discovery entered orbit safely and released the tracking and data relay satellite, which was its primary mission. Then the crew gave their remembrance of the *Challenger* crew.

Once in orbit, Mission Control contacted the crew and asked if they had seen anything pass by the window during launch. They confirmed that they had not. However, one of the engineering cameras viewed after the launch appeared to show something breaking off from the tip of the solid booster and flying down. Mission Control was concerned that this object may have struck the Orbiter's belly, causing damage, and so the crew were instructed to use the robot arm to look under the belly. What they saw shocked them. Something had extensively damaged the heat-resistant tiles that protect the Shuttle during re-entry.

> Hauck: *We were looking at this and saying: 'My god, that's a lot of damage.' And we saw one place looked like a tile was completely missing, but it looked to us like there was a lot of, lot of damage on the belly of this thing. We told Mission Control and they just kind of seemed blasé about it, like they were looking at the video, and they just didn't have a sense of urgency like I think we did, and expected them to have. It kind of baffled us. We said: 'Why are they not more concerned about this?' It was obvious to us there were probably hundreds of tiles that were damaged. And when we came back, it turned out that the video was such a poor quality with the Sun shining on those black tiles, it's hard to see things, is that they really couldn't see what we were seeing, and they saw a few scrapes and scratches and stuff and didn't think it was all that big of a deal, and I think everybody was shocked when the vehicle landed, and I think they ended up changing out like 700 heat tiles or something. It was a lot of heat tiles they had to change out that were damaged on that thing.*

Many Successful Satellite Launches and Space Station Missions

The Space Shuttle programme was never quite the same after the *Challenger* accident, even though it achieved a respectable, and safe, flight rate of about seven missions a year throughout the 1990s. Nevertheless, it launched the *Magellan* probe to Venus, the *Galileo* probe to Jupiter as well as the *Ulysses* probe into deep space. It also placed into orbit the great space observatories: the *Hubble Space Telescope*, the *Chandra X-ray Observatory* and the *Compton Gamma Ray Observatory*. It must be noted, however, that these remarkable satellites could have been launched on unmanned rockets – like the *Spitzer Infrared Space Observatory* was, on a *Delta 2* rocket. The Shuttle repaired the *Hubble Telescope* several times in orbit, greatly extending its life and scientific productivity. It flew nine missions to the *Mir* space station and many more to its successor, the *International Space Station*.

In October 1998 – 35 years after he became the first American to orbit the Earth – John Glenn returned to space on mission STS-95 becoming, at 77, the oldest person in space. This time he added another 134 orbits to his tally, ostensibly studying how an elderly person would adapt to zero gravity, as well as undertaking investigations into the ageing process, though it is questionable whether anything scientifically worthwhile was achieved.

The Space Shuttle has performed well, even with the to-be-expected close calls, but it will never be truly 'operational'. Using it to get into space involves the liberation of titanic amounts of energy that are too unforgiving of mistakes.

More Lucky Escapes

On the 95th Shuttle flight (and the 26th launch of *Columbia*), a small pin in the combustion chamber of one of the SSMEs broke loose during the early ascent phase. The pin was put there to 'repair' a previous fault. It struck the rocket's nozzle, puncturing its coolant jacket which was fed with liquid oxygen – the Shuttle's fuel. *Columbia* began to leak. It was not the only problem during launch. Five seconds after lift-off a short circuit disabled some of the controllers of two of its SSMEs. *Columbia*'s computers switched to the reserve controllers of the main engines, but it meant that two of its three rocket engines were one failure away from shutting down. Luckily, nothing else happened and the fuel leak was small. *Columbia* reached orbit safely, albeit a few miles lower than expected. But, as is so often the case, it could have been much worse.

A much more serious problem occurred during the launch of STS-112 – an *Atlantis* flight to the *International Space Station*. Only one set of the 'hold-down bolts' fired. Detecting the failure, the launch computer fired the backup system to release the bolts. It was vital that the bolts were free when the SRBs fired, otherwise *Atlantis* would have torn itself to pieces.

1988

29 September Shuttle STS-26 is launched, marking the first Shuttle flight since the *Challenger* disaster in January 1986

1989

4 May Shuttle mission STS-30 blasts off, and launches the Venus probe *Magellan*

18 October Shuttle *Atlantis* is launched on a mission that includes the deployment of the Jupiter probe *Galileo*

1990

24 April Shuttle *Discovery* is launched, carrying the *Hubble Space Telescope*

6 October Shuttle *Discovery* is launched, carrying the probe *Ulysses*

'It's serious. It's serious'

FIRE AND COLLISION

THE *MIR* SPACE STATION
1994–2001

The most difficult and dangerous space missions since the first Moon landings and Shuttle flights were the expeditions aboard the Russian *Mir* space station. The Shuttle–*Mir* programme was a collaboration between the United States and Russia during which Space Shuttles docked with *Mir* and US astronauts spent considerable periods on board the space station. During seven manned expeditions, Americans spent almost 1000 days in orbit. The collaboration saw the first American astronaut launched aboard a Soyuz and the first Russian cosmonaut flown on a Space Shuttle.

Rendezvous with *Mir*

The first phase of the collaboration began with the launch of Shuttle STS-60 in February 1994, with cosmonaut Sergei Krikalev on board *Discovery*. Although the Shuttle did not dock with *Mir*, there was a video link with the three cosmonauts on board. A year later the first female to command a Shuttle flight, Eileen Collins, flew *Discovery* alongside *Mir*. This time Russian cosmonaut Vladimir Titov was on board the Shuttle. A few months later astronaut Norm Thagard took off in a Soyuz capsule with Vladimir Dezhurov and Gennady Strekalov to visit *Mir*. He remained there for 115 days. They returned in July on Space Shuttle *Atlantis*, which carried out the first Shuttle–*Mir* docking after delivering two replacement cosmonauts. In 1995 the Shuttle delivered a new Docking Module and new solar arrays.

The following year the US maintained a significant presence on *Mir*. Shannon Lucid stayed for six weeks, up to September 1996. John Blaha took her place until the following January. Then Jerry Linengar stayed until May. Whilst on board he faced one of an astronaut's worst nightmares – a fire in space.

Fire!

There was a crew of six on board *Mir* at the time, February 1997. Usually they ate in shifts, but since that day was Russian Army Day they took a meal together. After the meal it was cosmonaut Aleksandr Lazutkin's

Space Shuttle *Atlantis* leaving the
Mir space station during the 1995
STS-71 mission.

routine task to reload a so-called oxygen candle in the Kvant Module. The oxygen generator uses three lithium perchlorate candles each day. When heated, they generate extra oxygen. As Lazutkin floated away after performing the task, he heard a hiss and then saw sparks quickly forming a flame. Reinhold Ewald, a European Space Agency (ESA) astronaut, saw it and shouted 'Pozhar,' meaning 'Fire'. Vasily Tsibilyev echoed his call: 'Pozhar. Pozhar.' Commander Korzun arrived with a fire extinguisher and the others grabbed one, too. The fire was growing. Black smoke was beginning to form. Lazutkin tried to switch off the device but there was no response. He threw a wet towel on it but it swiftly burned. Molten metal was dripping from it and the flame was now reaching towards *Mir*'s hull. If the hull became breached they would die in seconds; they all remembered *Soyuz 11*. Smoke stung Korzun's eyes as he flipped the fire extinguisher to 'foam' and depressed the button. Nothing happened. He shouted: 'Everyone to the oxygen masks, everyone stay in pairs.' He then ordered Lazutkin to prepare the ship – meaning one of the two Soyuz capsules docked at each end of *Mir*. The second Soyuz could not be reached through the fire. The alarm sounded. The flames were now 0.6 metres (2 ft) long and growing. Someone shouted: 'Where's Jerry?'

Jerry was almost asleep in the Spektr Module. He unstrapped himself, asking: 'Is it serious?' Ewald rushed into the Kristall Module to get more oxygen masks. 'It's serious. It's serious.' Tsibilyev and Linengar entered the Piroda Module to get more fire extinguishers but could not get them off the wall. So they moved to the Kvant 2 Module to get an extinguisher from there. Korzun was still fighting the fire and screaming for more extinguishers. By this time Lazutkin had reached the Soyuz and closed the hatch so that the smoke did not prevent their means of escape. By now the Kvant Module was dark and smoke filled, but Korzun knew he must go back and fight the flames, which now appeared as an ominous glow through the thick fumes. This time he flipped the extinguisher to 'water' and pointed the jet at the base of the hissing, spluttering flame. All too soon the extinguisher failed. 'I need more,' he shouted.

All *Mir* crews trained for an emergency evacuation. Lazutkin was preparing a Soyuz in case they had to abandon the space station. Valery Korzun, Aleksandr Kaleri and Reinhold Ewald's escape lay on the other side of the flame and smoke, in the shape of the second Soyuz craft. No one had reached it yet. If they had to escape, it was their only hope. Korzun kept the water jet on the flame. Thankfully it started shrinking. Through the smoke, the flame subsided. Slowly the smoke started to clear as *Mir*'s air conditioning took over. *Mir*'s wall was found to be badly scorched, but the hull was still intact. Linengar was succeeded by Michael Foale, who had to face his own crisis on *Mir*.

Unwelcome Trip to *Mir*

In the summer of 1995 Michael Foale had been a member of Shuttle mission STS-63, *Discovery*'s first rendezvous with *Mir*. About this time two of NASA's proposals for *Mir* astronauts, Scott Parazynski and Wendy Lawrence, were rejected as being unsuitable; one was too tall and the other was too short. Foale did not particularly want to go to *Mir*. He had in mind a trip to the *International Space Station* that was being planned. But while on a business trip to Star City he was told by someone that they had heard he would be coming to train there in a few weeks. 'Huh?' was his reply, before he became, in his own words, 'pretty angry'. Although he did not know it, NASA had agreed to send him to *Mir*. He arrived at *Mir* on 17 May

1997, on the Shuttle *Atlantis*. Later, when commenting on *Mir*, he said diplomatically: 'The condition of *Mir* is not the same as the *Space Station*.' Some Americans have said it smelled like a musty wine cellar, others that it smelled of sweat. There are cables and air ducts everywhere. Once on board, Foale was taken to see the site of the fire. No doubt he hoped that nothing like that would happen during his stay.

Coming in Too Fast

Six weeks into Foale's mission, it was decided to test a new control system for docking an unmanned *Progress* supply spacecraft. To do this they undocked a *Progress* already attached to *Mir* and backed it off, intending to use the new system to bring it back. When the *Progress* was about 4.4 miles (7 km) away, Tsibilyev took over manual control. He fired its thrusters to slow the approach. Although he was watching on a television screen, the *Progress* was approaching from below and he could see it against the Earth's clouds. Foale, holding the laser rangefinder, tried to see it from Kvant, but still could not make it out. Nobody could see the *Progress*, which was now 1.6 miles (2.5 km) distant.

It seemed to be coming in too slowly. It was only two minutes from the docking. It should have been visible by now. Tsibilyev put the brakes on, preparing for a hold point 400 metres (1312 ft) away. Where was it? Suddenly Lazutkin saw it emerge from behind a solar panel. It was big. Too big and too close. It was heading for base block. At 150 metres (492 ft) away Lazitkin shouted a warning: 'It's coming in too fast.' It was moving along the station, passing Kvant. Lazutkin shouted: 'It's moving past. Find Foale, get into the ship [the Soyuz].' Then Tsibilyev shouted: 'Oh Hell,' as he realized that a collision was inevitable.

The *Progress* crumpled a solar panel as it impacted. *Mir* shuddered. Then it struck the Spektr Module before rubbing along the side of *Mir* and moving away into space, tumbling end over end. The master alarm sounded. The hull had been breached, and *Mir* was leaking. Foale felt the pressure drop in his ears. Pressure was down to 600 millibars. At 540 millibars you can lose consciousness. The leak was in Spektr, and Lazutkin was gathering up the cables that entered the module and prevented its hatch from closing. The pressure was falling. They had to get the Soyuz craft ready. Things were getting frantic. They could not get Spektr's hatch to close, so Foale and Lazutkin went to search for a replacement hatch door.

After the collision, *Mir* was spinning at about a degree a second. Ground Control was concerned, asking them what was the spin rate. Foale moved quickly to the window and held his thumb against it, looking to the stars beyond to estimate a rough spin rate, which he relayed back down to the ground. They had to stop the spin, and that involved firing *Mir*'s manoeuvring thrusters in so-called 'blind mode'. They activated what they thought were the correct set of thrusters, and *Mir* ceased spinning. But they knew they had to start *Mir* slowly spinning in the correct manner again, so that the solar panels would remain pointing towards the Sun as it orbited the Earth. If they could not achieve this they would run out of power. But no one had been trained to do it. Then they lost all power and contact with the ground.

To get *Mir* spinning again they decided to use the working thrusters on one of the docked Soyuz craft. Foale suggested they should fire them in translation mode (usually used for moving the spacecraft) and not rotation mode (used for rotating it). Firing the thrusters in translation mode would have the greatest effect in turning *Mir*. But it was still risky, and in any case Foale would be the first to admit that he was not a *Mir* expert. If they were wrong or made a mistake, they could run down the Soyuz' fuel supply and jeopardize its use as an escape option. In the end they decided to fire a Soyuz thruster to see what effect it had on *Mir*'s orientation. But there was a problem, and it could be a big one.

Chasing the Sun

It was impossible to disconnect the Soyuz from *Mir*'s power supply, and then activate the Soyuz, if *Mir*'s power supply was dead. Fortunately, Tsibilyev had disconnected Soyuz just before *Mir*'s power failed, which meant he could operate its thrusters. Tsibilyev fired the thrusters and Foale observed the effect from the window. Slowly they worked out an effective procedure. To restore power to *Mir* they had to turn its solar panels towards the Sun, but it was orbital night. They looked at the darkened Earth. Where would the Sun rise? Then they saw some faint light streamers on the horizon. Foale said: 'Looks like we need to get the station over there.' By trial and error they tried to turn *Mir* in the right direction. They looked at the batteries; they were charging. It seemed that they had done it. Now they had to get the systems in the base block working – especially the carbon dioxide scrubbers, as carbon dioxide levels had been increasing.

After power and some semblance of order were established they organized a sleeping pattern. Someone had to be on watch at all times, but the rest needed to sleep. They would achieve nothing, and make mistakes, if they were too tired. They had to move dead batteries from the dead modules and charge them up in the base block, keeping a set of charged batteries in case they lost solar power again. After about 30 hours they had the base block working again, and after 48 hours, to their relief, the toilet, too.

A week after the collision – a week that was spent getting the power supply sorted – Ground Control suggested they undertake an 'internal spacewalk' into the Spektr Module to assess the damage. They could certainly do with the power from Spektr's solar panels, which was routed through the module. It seemed that only one of the four panels had been damaged by the collision. They needed the power from the other three. Engineers on the ground were working on a design for an adapter to be built into a hatch, which would allow electricity to be cabled out.

The crew prepared the spacesuits and worked on the procedures for the repair. In the meantime another *Progress* craft came up, this time docking successfully using the standard control system. It carried some much appreciated mail, as well as supplies and equipment for the internal spacewalk.

Into Another Spin

During the preparations it was assumed that two Russians would carry out the internal spacewalk. However, Tsibilyev developed a heart murmur, which resulted in Michael Foale being given the role of undertaking the spacewalk in a Russian spacesuit. Before the spacewalk could proceed, about 100 cables running through the base block to Kvant 2 had to be disconnected so that the internal hatches could be closed. Some of the cables were connected to the gyroscopes that controlled *Mir*'s orientation. Only two days before the planned spacewalk one of the cables was disconnected out of sequence, causing *Mir* to go into a big tumble as the gyroscopes spun down. Technically, it was a more serious power-down situation than the one that followed the collision.

A Rushed Choice of Crew

At this point mission controllers realized that they had been pushing the crew too hard, and they were also worried about Tsibilyev's mental condition. They decided that the internal spacewalk would be performed by the next crew, whose arrival was brought forward a week, shortening the crew overlap time and cancelling the flight of an ESA astronaut due to visit *Mir*.

The next crew, *Mir Expedition 24*, arrived, composed of Anatoli Solovyov and Pavel Vinogradov. In subsequent weeks they performed the internal spacewalk, although they were unable to completely restore the power. Foale then made an exterior spacewalk to assess damage. The leak was never fixed.

The night before they left *Mir* it is reported that Tsibilyev and Lazutkin stayed up late autographing and stamping stationery, photographs and letters to take back home to sell. Foale estimated that they only had about two hours sleep and that it was an irresponsible thing to do.

The End of *Mir*

After the emergencies, the US Congress considered whether the Americans should abandon the *Mir* collaboration programme out of concern for the astronauts' safety, but NASA Administrator Dan Goldin chose to carry on. In June 1998 the final US-*Mir* astronaut, Andy Thomas, left the space station aboard the Shuttle *Discovery*.

The 39th, and final, manned mission to *Mir* was *Soyuz TM-30*, launched in April 2000. Given sufficient resources *Mir* could perhaps have been patched up and helped to continue, but Russia's commitment to the *International Space Station* and its limited resources meant that it had to be cancelled. *Mir* was brought back to Earth on 23 March 2001, near Fiji.

1994

6 February Sergei Krikalev becomes the first cosmonaut to fly on the Space Shuttle (STS-60), when it approaches the *Mir* Space Station

1995

16 March Norman Thagard becomes the first astronaut to be welcomed aboard *Mir*

May *Mir* is reconfigured to receive the US Space Shuttle

29 June Space Shuttle docks with *Mir*, becoming the biggest man-made satellite to orbit the Earth

1996

22 March The Shuttle STS-76 blasts off for *Mir*. Among the crew is astronaut Shannon Lucid, the first woman to live on the station

1997

23 February Crew of *Mir* experience a severe fire on board

25 June A Soviet *Progress* supply craft collides with *Mir* while docking, causing *Mir* to lose cabin pressure and spin out of control

2000

4 April The final manned mission to *Mir*, *Soyuz TM-30*, is launched

2001

23 March Parts of the *Mir* space station are visible as they burn out on entry into the Earth's atmosphere

'Off-scale low'

END OF AN ERA

STS-107 – THE FINAL FLIGHT OF COLUMBIA
2003

Columbia began its return to Earth at the completion of STS-107 on 1 February 2003. It was its 28th flight, and its last. It had been launched almost 16 days earlier on a wide-ranging science mission to study the Earth's atmosphere and microgravity. About 82 seconds after launch, when *Columbia* was at an altitude of 20,116 metres (66,000 ft), a suitcase-sized piece of thermal insulation foam broke off from the external tank's bipod foam ramp section and struck the leading edge of the spacecraft's left wing.

The impact with the large piece of insulation material is believed to have created a 15–25 cm (6–10 in) hole in the Space Shuttle's wing. It was not the first time that foam had fallen from this area; it had happened on at least four previous Shuttle flights. As *Columbia* entered its second orbit around the Earth, mission controllers reviewed video taken of the launch and concluded that it was nothing unusual. The following day higher resolution video revealed that the foam had struck the wing, but the extent of the damage, if any, was not possible to determine. Engineers wanted to use secret spy satellites to take a look at *Columbia* in orbit, but NASA officials deemed this unnecessary. Another engineer was so concerned that he requested that an astronaut visually inspect the area, but again his request was declined. In total, three requests for in-orbit imagery of *Columbia* were turned down. The subsequent accident report revealed that the concerned engineers found themselves in the position of having to prove that the Shuttle was unsafe – a reversal of the usual requirement of proving that a situation is safe.

Countdown to Disaster

2.30 a.m. Eastern Time: the Entry Flight Control Team began their shift in Houston's Mission Control Center. At handover, there were no issues raised that were out of the ordinary. They went through their checklists; it was a re-entry like any other. The weather at the Kennedy Space Center in Florida was good.

8.00 a.m.: Flight Director Larry Cain polled Mission Control for a go/no go for re-entry. A few minutes later the capcom notified the crew that they were go for the de-orbit burn. Commander Rick Husband and Pilot William McCool fired *Columbia*'s two OMS engines. Flying upside down and tail first over the Indian Ocean at 175 miles (282 km) altitude, they were on their 255th orbit. The 2-minute, 38-second burn slowed them sufficiently to begin their entry into the atmosphere. They were now committed. There was no turning back.

The official crew photograph from mission STS-107
on the doomed Space Shuttle *Columbia*.

After the burn Husband turned *Columbia* around and pitched its nose up so that the heat-resistant tiles on its belly would face the brunt of re-entry friction.

8.44 a.m.: *Columbia* reached the so-called Entry Interface at 121,920 metres (400,000 ft) – the point at which the first signs of the atmosphere are evident. They were passing over the Pacific Ocean.

8.48 a.m.: a sensor on the left wing's leading edge showed strains higher than those recorded on previous Space Shuttle re-entries, but the data was stored on the onboard flight recorder and not transmitted to the ground controllers. For them, the re-entry was normal, but that was about to change.

8.49 and 53 seconds a.m.: *Columbia*, travelling at Mach 24.5, made a pre-planned turn to the right to adjust its rate of descent and heating. Sixty seconds later it entered the 10-minute-long period of peak heating, when thermal stresses are at their greatest. It was nearing the Californian coastline. Wing leading edge temperatures now approached 1450 °C. The speed had dropped slightly to Mach 23. *Columbia*'s altitude was 70,400 metres (231,000 ft). Observers reported sighting debris being shed from *Columbia*. Flares were seen in the superheated air surrounding the Space Shuttle.

8.54 and 24 seconds a.m.: the Maintenance, Mechanical and Crew Systems Officer informed the Flight Director that four hydraulic sensors in the left wing were 'off-scale low', indicating that the sensors had failed in some way. As Columbia crossed into Nevada airspace at 69,190 metres (227,000 ft), witnesses reported seeing the start of a series of bright explosions. Wing leading edge temperatures were now at their maximum – 1650 °C.

'Roger, uh, bu...'

RICK HUSBAND

8.56 and 30 seconds a.m.: *Columbia* was over Arizona and starting another turn.

8.59 and 15 seconds a.m.: pressure readings were lost from both left wing landing tyres. At this time Rick Husband had been trying to say something. The Flight Director told the capcom that they did not understand *Columbia*'s last transmission. Seventeen seconds later Husband was heard to say: 'Roger, uh, bu...' He was cut off mid-word. Observers could see that *Columbia* was breaking up, but as yet Mission Control saw no sign of any serious trouble.

9.05 a.m.: observers saw smoke trails and debris falling in the sky over Texas.

9.12 a.m.: the Flight Director declared a contingency, which meant the loss of the vehicle. Search and Rescue teams were alerted. He issued an instruction to lock the doors. No one was allowed in or out. Flight controllers were told to preserve all the mission data for the investigation that would follow. They all knew the crew was dead.

The Last Moments

Hot gasses started entering through a breach in the left wing at 8.44 a.m., although the crew were unaware until several minutes later. They survived for a minute after their last communication with Mission Control, many minutes after it was obvious to them that something was going seriously wrong. Just like in the *Challenger* accident, the crew cabin held together. A data recorder called the orbital experiment support systems recorder was recovered from Hemphill in Texas. It had continued to operate until 18 seconds after 9.00 a.m. – almost a minute after Rick Husband's incomplete last communication. Following this last voice transmission data was lost for 25 seconds, then resumed for two seconds. In those two seconds, data received on the ground indicated that conditions in the cabin were benign and that, possibly, the autopilot had been disengaged.

The cabin remained intact as *Columbia* disintegrated. Rick Husband would have worked the hand controller trying to regain some control but the rapid tumbling would have told him, and the rest of the crew, that there was no Shuttle to control. We do not know how long they survived before the cabin itself fragmented about 30 seconds later, but certainly they had time to understand their fate. The investigation concluded that they died of 'blunt trauma and hypoxia with no evidence of lethal injury from thermal effects'. Three of the crew were known not to be wearing gloves at the time of the accident and one was not wearing a helmet.

Searching for Clues

There were more than 2000 debris fields, including human remains, scattered across Texas, Louisiana and Arkansas. A small culture of *Caenorhabditis elegans* worms were found still alive in petri dishes enclosed in aluminium containers. Among the wreckage was a 13-minute video made by the astronauts during the start of re-entry. There is no indication of any problem as the crew go through re-entry procedures, joking with each other as they do so. The tape ends four minutes before the accident. That portion of the video was destroyed in the crash.

An investigation board was established that included Sally Ride, who had also served on the Rogers Commission investigating the *Challenger* accident. She noted the similarities between the two accidents and asked why the Shuttle was allowed to fly with known problems that were, eventually, catastrophic.

Throughout the stifled debate about the true extent of the damage caused to *Columbia*, many NASA managers were convinced nothing could have been done even if the extent of the damage had been known. But that is to ignore NASA's history. When the lives of astronauts are at stake, remarkable things can be achieved. A spacewalking in-orbit repair might have been possible, or a second Shuttle rushed into space to transfer the crew. Believing that there was no point in further investigation or action ran contrary to the 'failure is not an option' attitude of previous space programmes. *Columbia*'s last crew deserved better.

Henceforth, all Shuttle missions save one – the final refurbishment flight to the *Hubble Space Telescope* – were to go to the *International Space Station* where the Shuttle's tiles and wings could be inspected for problems, and if needed the crew could find a safe haven in the space station.

2003

16 January Space Shuttle *Columbia* STS-107 is launched from Kennedy Space Center to undertake a series of scientific investigations

1 February *Columbia* disintegrates on its re-entry, killing all seven crew members

Winding Up the Programme

In July 2008 NASA set the dates for the final flights of the Space Shuttle before it is retired in 2010. Seven missions will go to the *International Space Station* and one to the *Hubble Space Telescope*. If all goes well, the last flight for *Atlantis* will be in February 2010; *Discovery*'s last flight will be in April; and the final flight of all, *Endeavour*, will be in May. Nevertheless, there has been some recent debate about the wisdom of relying for several years on Russian Soyuz spacecraft as the only manned access to the *International Space Station*. In the light of this, a campaign is gaining headway to extend the Shuttle's operating lifetime until 2015.

Despite this, attention is turning to its successors – the *Ares* rockets and the Orion crew vehicle.

Largest structure in space

LIVING IN ORBIT

THE *INTERNATIONAL SPACE STATION*
1998–

Assembly of the *International Space Station* **(ISS) began in 1998, although the project had been off and on for years. Wernher von Braun would not recognize it. In appearance, it is an angular, roughly H-shaped structure consisting of cylindrical modules flanked by arrays of huge, oblong solar panels. The *ISS* is a research facility, carrying out work in fields such as biology, astronomy, meteorology and physics.**

The largest structure ever built in space, the *ISS* completes almost 16 orbits a day 217 miles (350 km) above our heads and is easily visible to the naked eye as a bright star-like object moving swiftly across the sky. Anyone with an Internet connection can see and hear what is going on there in real time.

The *ISS* has been inhabited continuously since its first crew arrived in November 2000. At present, three can live aboard it long-term. It has been visited by astronauts from 16 countries, and it has also played host to five space tourists – wealthy individuals who purchase training and transport from the Russians for sums upwards of $10 million. One resident, Yuri Malenchenko, got married on board the *ISS* in 2003. His bride was in Texas, and he was flying over New Zealand at the time.

The *ISS* must be counted a success, having suffered only minor problems of the sort that are likely to occur when living and working in space. There has been the odd smoke incident, torn solar panels, faulty bearings and a crashed computer, but nothing so far on the scale of previous missions and previous space stations. For the most part, the station operates relatively quietly and efficiently.

The Soyuz 'Shuttle'

In February 2003, after the *Columbia* disaster and the suspension of Shuttle flights, two American astronauts found themselves stranded on the *ISS* with no ride home. Ken Bowersox, on his fifth space mission, and Don Pettit on his first, had expected to return in March 2003 on the Shuttle *Atlantis*. But

---

1998

20 November A Russian *Proton* rocket launches with the first stage of the *International Space Station*

4 December Shuttle *Endeavour* launches, carrying the second component of the *ISS*

2000

July Addition of the Service Module, renders the *ISS* habitable

2003

3 May Two American astronauts, stranded on the *ISS* following the grounding of the Space Shuttle, are ferried back to Earth aboard a Russian Soyuz spacecraft

then Jefferson Howell, director of the Johnson Space Center, spoke to them on the radio. 'I have some bad news,' he told them, 'we've lost the vehicle.' Instead the astronauts came back in *Soyuz TMA-1* in May.

All crew exchanges between February 2003 and July 2006 were carried out by Soyuz craft. The basic Soyuz design is over 40 years old – Korolev's ideas live on. It is still designed for small humans, although a little taller than in the days of Yuri Gagarin. Pettit and Bowersox were the first American astronauts to return home on a foreign vessel and the first American astronauts since 1975 to return in a capsule. After suiting up, Soyuz passengers lie on their backs, with their knees pulled up close to their chests. After a long checklist, the commander presses a single button, once, and that is all. The ride home is usually automatic.

The Chinese in Space

In October 2003 China sent its first astronaut into orbit using a Shenzhou spacecraft – a design that owes much to the Soyuz – becoming only the third nation to launch its own astronauts. Yang Liwei was 38 at the time. He did not enjoy a comfortable ride into space; the rockets' vibrations during ascent were hard to endure. He stayed in space for 21 hours. In October 2005 two Chinese astronauts spent five days in space. More will follow. It is thought that the Chinese want to put a human on the Moon. That would be easier today with advanced computers, better materials and the lessons of history to help. The Chinese have also said they want to carry out spacewalking and docking by 2012, and a moonwalk by 2024.

The *International Space Station* photographed by the
Space Shuttle *Atlantis* in 2002.

'The flight was spectacular'

A TICKET TO RIDE

SPACESHIPONE
2004

In June 2004 the first privately funded spacecraft went into space. It had been designed and developed by Scaled Composites, an adventurous and well-respected aerospace design company that had used the genius of Bert Rutan to conceive many remarkable aircraft.

SpaceShipOne was taken to an altitude of 15,240 metres (50,000 feet) from where it detached from its carrier aircraft and fired a so-called hybrid rocket motor to take it straight up to an altitude of just over 62 miles (100 km), officially entering space. The pilot, Mike Melvill, 64, became the world's first commercial astronaut. After touchdown he told the cheering crowd:

> *The flight was spectacular. Looking out that window, seeing the white clouds in the LA Basin, it looked like snow on the ground.*

During the dramatic flight, the craft had surprised him with its 'little victory roll', and he had to shut down the engines 11 seconds prematurely:

> *Did I plan the roll? I'd like to say I did but I didn't. You're extremely busy at that point. Probably I stepped on something too quickly and caused the roll but it's nice to do a roll at the top of the climb.*

Two others, Peter Siebold and Brian Binnie, have so far also flown *SpaceShipOne* to space. Others will follow soon, when its successor *SpaceShipTwo* carries passengers for $150,00 each on a suborbital flight from late 2009. Some 48 years after Alan Shepard rode a *Redstone* rocket on a suborbital flight to become the first American in space, it is now possible to buy a ticket to do the same thing. By August 2008, a total of 484 people had become astronauts. Soon there will be hundreds more, then thousands.

A New Breed

In late 2007 NASA issued a request for applications for a new class of astronaut. For the first time in decades the successful candidates will not study to fly aboard the Space Shuttle. Instead they will train, for the first time since the late 1960s, to walk upon the Moon.

SpaceShipOne gliding after
having been released at high
altitude by its carrier craft,
White Knight.

To paraphrase the words of the American poet Walt Whitman, they will venture where mariners have
not yet dared to go, and risk the ship, themselves, and all to get there. Like those before them, they will
sometimes cheer and sometimes cry. Somewhere, a young child is growing up who will not only become
captivated by our future voyages to the Moon but will also cast their imagination even further. At this very
moment, the first person to set foot upon Mars is dreaming of astronauts.

Index

Picture credits

2-3 Great Images in NASA; 5 Great Images in NASA; 7 Library of Congress / Science Photo Library; 8 NASA Headquarters - Greatest Images of NASA/Asif Siddiqi; 12 Detlev van Ravenswaay / Science Photo Library; 16 Shutterstock/Dmitry Bodrov; 19 Great Images in NASA; 23 NASA; 25 Ria Novosti /Science Photo Library; 27 NASA Langley Research Center; 37 NASA Headquarters - Greatest Images of NASA/Asif Siddiqi; 39 Ria Novosti /Science Photo Library; 43 Great Images in NASA; 48 Great Images in NASA; 53 Ria Novosti /Science Photo Library; 56 Great Images in NASA; 58 NASA Headquarters - Greatest Images of NASA/Asif Siddiqi; 61 Great Images in NASA; 64 NASA Headquarters - Greatest Images of NASA/Asif Siddiqi; 69 Great Images in NASA; 71 Great Images in NASA; 75 Great Images in NASA; 80 Great Images in NASA; 83 Great Images in NASA; 84 Great Images in NASA; 88 Ria Novosti /Science Photo Library; 93 NASA Headquarters - Greatest Images of NASA/Asif Siddiqi; 95 Great Images in NASA; 98 Soviet Space Program/ USGS/Michael Benson, Kineton Pictures; 101 Great Images in NASA; 105 NASA Headquarters - Greatest Images of NASA; 107 NASA Headquarters - Greatest Images of NASA/Asif Siddiqi; 111 Great Images in NASA; 115 NASA Headquarters - Greatest Images of NASA; 119 Great Images in NASA; 121 NASA Marshall Space Flight Center; 122 Great Images in NASA; 126 Great Images in NASA; 129 Great Images in NASA; 133 Great Images in NASA; 139 NASA; 143 Ria Novosti /Science Photo Library; 147 Great Images in NASA; 150 Great Images in NASA; 153 Great Images in NASA; 155 Great Images in NASA; 159 NASA/ Science Photo Library; 162 Great Images in NASA; 167 Master Sgt. Dave Casey/Department of Defense; 169 Great Images in NASA; 173 Great Images in NASA; 179 Ria Novosti /Science Photo Library; 183 Great Images in NASA; 187 Great Images in NASA; 191 Great Images in NASA; 197 Great Images in NASA; 201 NASA/Space Shuttle Gallery; 203 Scaled Composites/ Science Photo Library.

Quercus Publishing has made every effort to trace copyright holders of the pictures used in this book. Anyone having claims to ownership not identified above is invited to contact Quercus Publishing.

Audio CD translations

The following provides a translation for each of the five original Russian recordings featured on the accompanying audio CD:

Track 3. Yuri Gagarin speaks to ground control during launch. 12 April 1961:

'My condition is excellent. All is well. I hear you loud and clear.'

Track 8. Valentina Tereshkova speaks in orbit, calling ground-tracking station. 16 June 1963:

'I feel fine. All is well. Please, please respond.'

Track 9. Vladimir Komarov speaks in orbit. 12 October 1964:

'*Zarya 8*, this is Ruby. Do you read?'

Track 10. Ground control calls crew of *Voskhod 2*. 18 March 1965:

'Almaz, Almaz. This is *Vjezna 3*. The circuit for the automatic descent system has been switched off and the system to control descent manually has been activated.'

Crew: 'We acknowledge.'

(Almaz was the *Voskhod* call sign)

Track 17. Komarov speaks onboard the spacecraft *Soyuz 1*, praising the Communist Party. 23 April 1967:

'For the benefit of the people of the fatherland and for the whole of humanity on the famous way to Communism.'

For my late father-in-law Bernard Carey. With thanks.

Quercus Publishing Plc
21 Bloomsbury Square
London
WC1A 2NS

First published in 2009

Copyright © David Whitehouse 2009

The moral right of David Whitehouse to be identified as the author of this work has been asserted in accordance with the Copyright, Design and Patents Act, 1988.

A CIP catalogue record for this book is available from the British Library

Printed case edition with CD: 978 1 84916 067 4

Paperback edition with CD: 978 1 84866 036 6

Printed and bound in China

10 9 8 7 6 5 4 3 2 1

Designed and edited by BCS Publishing Limited, Oxford.